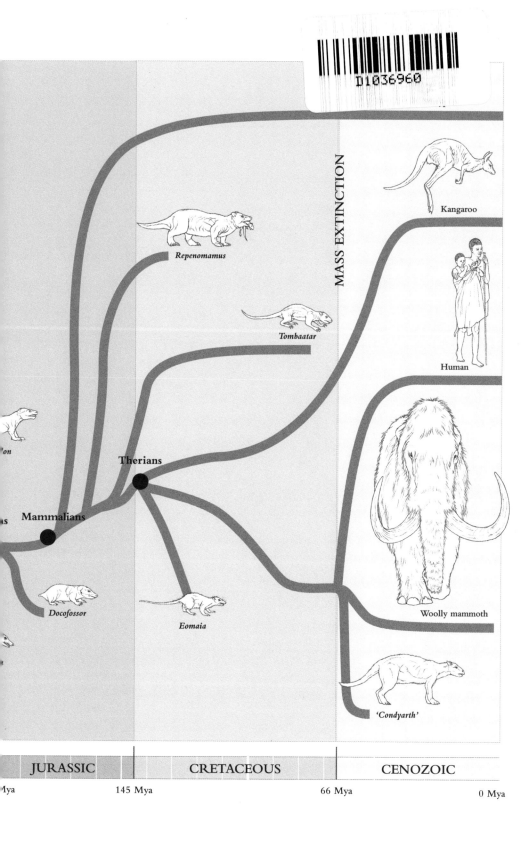

Kangaroo

Human

Repenomamus

Tombaatar

Therians

Mammalians

Docofossor

Eomaia

Woolly mammoth

'Condyarth'

MASS EXTINCTION

JURASSIC	CRETACEOUS	CENOZOIC

145 Mya   66 Mya   0 Mya

BEASTS BEFORE US

BEASTS BEFORE US

*The Untold Story of Mammal
Origins and Evolution*

Elsa Panciroli

BLOOMSBURY SIGMA
LONDON · OXFORD · NEW YORK · NEW DELHI · SYDNEY

BLOOMSBURY SIGMA
Bloomsbury Publishing Plc
50 Bedford Square, London, WC1B 3DP, UK
29 Earlsfort Terrace, Dublin 2, Ireland

BLOOMSBURY, BLOOMSBURY SIGMA and the Bloomsbury Sigma logo
are trademarks of Bloomsbury Publishing Plc

First published in the United Kingdom in 2021

Photo credits (t = top, b = bottom, l = left, r = right, c = centre)

Colour section: P. 1: © Elsa Panciroli. P. 2: © Chicago Field Museum, Chicago, USA (t); ©
Royal Tyrrell Museum of Palaeontology, Drumheller, Alberta (b). P. 3: Chicago Field Museum,
Chicago, USA (t); © Biodiversity Heritage Library (b). P. 4: © Institute of Palaeobiology, Poland
(t, c, b). P. 5: © Zhe Xi Luo / University of Chicago (t, cl, cr); © Oxford University Museum
of Natural History (b). P. 6: © Julien Benoit (tl, tr); © Oxford University Museum of Natural
History (c); © Elsa Panciroli (b). P. 7: © Matt Humpage; © Zhe Xi Luo / University of Chicago
(b). P. 8: © Elsa Panciroli (t); © Michael Waldman (c, b).

A catalogue record for this book is available from the British Library

Library of Congress Cataloguing-in-Publication data has been applied for

ISBN: HB: 978-1-4729-8382-4; eBook: 978-1-4729-8397-8

2 4 6 8 10 9 7 5 3

Typeset by Deanta Global Publishing Services, Chennai, India
Printed and bound in Great Britain by CPI Group (UK) Ltd, Croydon CR0 4YY

Endpaper illustrations by Marc Dando

Chapter illustrations by April Neander

Bloomsbury Sigma, Book Sixty-five

To find out more about our authors and books visit www.bloomsbury.com
and sign up for our newsletters

To my favourite mammals of all, my family.

Contents

Introduction

If you are reading this book, I probably don't have to convince you that fossils are interesting – but are they *useful*? Is palaeontology the dinosaur of science? Can old bones provide information relevant to the modern world?

When I began my palaeontology Masters degree, we started with a lecture in which my professor – himself the author of multiple books on the topic – opened with the famous quote attributed to Nobel Prize-winning scientist Ernest Rutherford: 'All science is either physics or stamp-collecting.' No prizes for guessing what *he* studied. My professor spent the lecture explaining the many reasons why palaeontology was no mere hobby. I remember being puzzled; it hadn't occurred to me that anyone would ever think this way. Surely all science is created equal?

There is a perception that the study of palaeontology is just the study of a bunch of dead things. Apart from entertaining children, it can appear at first glance as though there's little to it but describing dusty bones, hacked from the Badlands by men in wide-brimmed hats. Many people believe it is all about dinosaurs. 'I know a five-year-old who would love to meet you!' people say to me. That's lovely, I reply, but do you know any *forty*-five-year-olds I could speak to?

Fossils are unique for the insights they provide into the deepest origins of life on Earth. They yield answers to questions that molecules can never address. At first blush, bones and teeth betray the presence and absence of organisms – when did certain groups appear and disappear (or at least, what is the earliest or latest we know they certainly lived), and where did they inhabit? Next, they give us information about taxonomy, the study of how different groups are related to one another, which provides a framework for our understanding of all animal relationships. Fossils tell us how diverse animals and plants were at different times in the past, and the changing environments they inhabited. We can then use them to chart the outlandish journey evolution has made: natural selection's incremental tip-toeing and occasional leaps. We can also chart fluctuating climate and atmosphere, ocean acidity and temperature, and the function and fecundity of ecosystems. Taken together, this information allows us to identify evolutionary patterns taking place across geological time on a global scale. As we face what has been rightfully named the sixth mass extinction – a human-made natural disaster – it has never been more vital for scientists to know how life has responded to extinction events in the past, and most crucially, how it has recovered.

But going beyond that, these old bones can tell us how extinct animals once *moved*. Did they run, hop or slither their way across the face of the Earth? And what was their role in their environment, their *ecology*? This kind of information has been inferred for centuries via observation of skeletons, but we can now compute enormous datasets of bone shapes and test those hypotheses about form and function mathematically. We can even nick a few tricks from engineering, taking methods used to test the strength of building materials and applying them to fossils to understand their capabilities. The methods and results of these transformative new analyses have applications for human and veterinary medicine, for conservation and ecology, and of course for our knowledge of extinct life itself.

If you still have that Indiana Jones* picture of the palaeontologist in your mind – or even Ross from *Friends* – you couldn't be more wrong. Forget whips, what real palaeontologists carry with them is their laptop. They may need a hat for occasional fieldwork, but most would do better investing in a comfy office chair because they'll be spending most of their lives in front of a computer. Coding is a vital tool in the kit, allowing them to gather enormous volumes of information and statistically analyse it. They type code like most of us type text messages. They fetch CT scans like ordinary people grabbing a takeaway coffee.

We are now seeing a radical transformation in the science of extinct life. That is part of the reason that I am writing this book. The use of statistical methods to analyse big data and the routine CT scanning of fossils have opened up entirely new fields of research. Our knowledge

---

* Who was of course not a palaeontologist at all, but a whip-bearing, hat-wearing archaeologist – and a really bad one at that.

has accordingly exploded out of the boundaries to which pen, paper and a keen eye had previously confined it. The role of women is finally being recognised, both in the past and in the present. There is still a long way to go to address the diversity balance in palaeontology (as in other sciences) – we're still a little too white and Western. But increasingly, the countries where the most ground-breaking discoveries are being made are not Europe and America, but China, Madagascar, South Africa, Argentina, Brazil. Researchers from those countries are studying their own fossil heritage, rather than seeing it removed to museums abroad, as was once habitually the case.

The main reason I'm writing this, however, is to set the mammal story straight. You might think you know where mammals come from, but I reckon you'll be surprised. If you thought it all began with the extinction of the non-bird dinosaurs, think again. If you believed mammals – we fur-ball milk-givers – merely scooted underfoot like terrified snacks for most of the time of the dinosaurs, you are dead wrong. If you have always repeated the tale that mammals come from reptile stock, wash your mouth out. We mammals are a lineage all our own. Our branch stretches away from the others, tied through time by our anatomy and physiology to the first backboned animals on land. Long ago we struck out in a flowerless Eden, and we made good.

I want to show you the parts of our mammal journey you've never seen before. It's impossible for me to introduce you to every animal, group, place and researcher in the history of mammals – neither would you enjoy it if I did. Instead, I've selected a smorgasbord of the tastiest morsels. I'll showcase astonishing fossil discoveries, the key twists and turns in mammal evolution, and a few of the many talented researchers who have transformed our knowledge. There are some stand-out characters you should meet, and localities to tour. I especially want to invite you to Scotland, and show you how our discoveries among the seaweed of the Western Isles fit into a global picture of mammal development, linking to mind-blowing new fossils recovered from places like China and South Africa.

I'm going to take you on a completely new journey through mammal evolution.

This is the prequel. This is the *untold* story.

CHAPTER ONE

Isle of Mists and Lagoons

Cha b' ann gus a-rithist a thuig mi na trì mionaidean sin a bhi nan triall-farraige, bho sheann chreagan Eòrpach gu fìor chreagan àrsaidh Aimeireaga. An dà thìr mhòr air am brùthadh ri guailibh a-chèile, gus gleann a dhèanamh de chuan. Air a bhith air taobh thall a' chaolais, taobh na Morbhairne, taobh Àird nam Murchan. Bha mi, a-nis, air an tìr eile.

It was only later that I understood the three-minute journey for the sea-pilgrimage it truly was, from the old rocks of Europe to the ancient ones of America. The two continents pressed shoulder-to-shoulder, making a glen of an ocean. Standing now on the other side of the narrows, the Morvern – the Ardnamurchan side, I was now in another land.

Màiri Anna NicUalraig (Mary Ann Kennedy), *Caolas* (*Narrows*)

I pushed my bag as far as possible under the boulder, where a patch of grass escaped the downpour. My own body was too large to fit into this dry haven, so I wedged myself into the leeside of the rock. I pulled up both of my hoods – the one on my outer waterproof jacket, and the one on my secondary waterproof overcoat – to cover my sodden grey hat. Taking off my gloves I wrung them out onto the limestone pavement, watching the water cascade into the cracks sloping to the sea.

It was the last day of fieldwork on the Isle of Skye, the largest of the islands in the Inner Hebrides, a scatter of emeralds along the west coast of Scotland. The Gaelic name for this place is Eilean a' Cheò, the Island of Mist.[1] That day, the clouds hung so low that they swallowed the Cuillin mountain range completely. I could only see their stone-cold feet, thrust moodily into the choppy waters of Loch Scavaig.

Below me on the foreshore, two men were whacking the rock with big hammers. Flakes of stone zoomed in every direction like stray bullets. That was why I was sitting up there, to dodge the flying shards. One man had his long legs tucked under him in a rising pool of water, as the rain poured over his thin waterproof layers. The other was standing bent at the waist, his Gore-Tex-clad figure shimmering with a film of moisture as he pounded stone.

These are my colleagues, Roger Benson and Stig Walsh. Benson is tall and scarecrow-ish like me, his dark hair perennially in need of cutting. He has a restless mind and endless capacity for discussing science, punctuated with an off-beat sense of humour. A physics student turned palaeontologist, he began his career working on *Megalosaurus*, the first dinosaur ever named, but an interest in the big questions about animal evolution have focused his laser-attention on almost every vertebrate group, from fish to birds to mammals. Walsh is quite singular. A specialist in the brain and sensory evolution, he is also a hill-runner – famously in bare feet, though he's always shod when setting off from our office each lunchtime to speed around Arthur's Seat strapped to a GPS pedometer. There are so many stories to tell about his lifetime of misadventures, strange jobs and surprising hobbies, that I briefly considered abandoning this book to pen his biography instead. His random anecdotes begin with ruminations like, 'The inside of a nuclear reactor is not at all like you would expect.' He's the only person I know of who has been nibbled by snails in his sleep.

This was the last day of our annual field work in southern Skye. Most days we'd been lucky, not only returning to our rented cottage each evening with new vertebrate fossils to study, but sporting suntans. On our last day the weather had turned. We'd gone back to the shore to 'dress' the week's extraction sites. We knock the edges from our rock-saw cuts, leaving a surface that given some time, slightly acidic rainwater and the fervent colonisation of barnacles and algae, will be almost invisible in a few years. This approach means we have a minimal long-term impact on the area, an important stipulation in our permits from Scottish Natural Heritage (now NatureScot). This coast is a site of special scientific interest (SSSI). We are careful only to take what are likely to be significant finds. Our work is timed to avoid the breeding season for the many wild birds that frequent the area, including magnificent golden eagles. We care about more than just the dead.

I remember the field season when I met everyone for the first time. I entered the cosy rented cottage to be greeted by my new team members, who were sitting round the glowing log fire. The air was musty with damp clothing and wood smoke. Dinner had been set aside for my arrival, and I ate it from a bowl on my lap, curled up on a tattered leather armchair.

We spent each evening drinking peaty whisky and cheap wine, discussing how to calculate patterns of tetrapod diversity through time – palaeontologists don't half know how to party. Our group is multi-institutional: Walsh is Senior Curator of Palaeobiology at the National Museum of Scotland and associate researcher at the University of Edinburgh. He was also my PhD supervisor. Benson travelled north each year from his lab in the Earth Sciences department at Oxford University. Our other core team-mate was Richard Butler from the University of Birmingham, a red-headed giant never without bird-watching binoculars. Students and post-doctoral researchers also joined us each year. Benson and Walsh had been visiting Skye for six years when I began working with them. After a decade of fieldwork, they were convinced each year that there would be nothing more to collect, yet the discoveries continued as we grew to know the stone intimately, and teased out its secrets.

Benson, Butler and I are all over six feet tall, perhaps giving our group the greatest median height of any palaeontological collaboration in the United Kingdom. We are a team of giants in search of dwarves – the smallest vertebrate animal fossils hidden in Skye's grey limestone. Mammals and salamanders, lizards, turtles and pterosaurs. A dinosaur or two, perhaps. These were the inhabitants of a 166 million-year-old freshwater ecosystem, when Skye was a network of warm lagoons.

We drank our drams and made plans to hike out to the rocks in the morning.

My first trip to Skye had also been swallowed by rainclouds. I was 12 years old and my family packed themselves into the car and set off for the coast. We lived on the shores of Loch Ness at this time, and the road to Skye led through some of the most impressive mountains in Scotland. None of them were visible of course, from our sub-nebular position in the bottom of the ice-carved glens.

My parents made a big fuss about driving over the newly constructed bridge from Kyle of Lochalsh to Kyleakin, connecting the mainland to the island. They reminisced about the old car ferry – which now only runs for curious tourists and nostalgic locals – and how different life would be for the inhabitants of the island. This was discussed with both appreciation and regret, as most improvements to transport links in Scotland are.

When our tyres hit Skye's tarmac it was like diving under a wave. The rain didn't fall, it just existed everywhere at once. We saw almost nothing on our day-trip except pervasive russet gloom, with a soundtrack of frantically beating windscreen wipers. Dad, a mountaineer, talked about the incredible peaks of the Cuillin mountain range, which he swore were *right there*, just beyond the clouds. I gazed absent-mindedly out of the window at the murk.

That was when I saw it, perched on a fence post by the roadside: a hook-beaked silhouette. Reaper of rabbits. Giant wings folded like closed stable doors over its oak-brown back. '*I saw a golden eagle!*' I exclaimed to my parents, pressing my nose to the window. The car had shot past it before anyone else could see.

Mum asked if it wasn't just a buzzard. No, it was an eagle all right: the first I'd ever seen, dwarfing those everyday buzzards. A magnification of raptordom. A regal giant of the bird world.

My first dinosaur on Skye.

Twenty years later I stood on the shoreline of the Trotternish Peninsula in northern Skye, with my big green hiking-boot in a Jurassic dinosaur's footstep. I was at Rubha nam Brathairean, near Staffin: the most remote[*] dinosaur footprint site in the UK. I had walked down to the shore to prospect for fossils with a team from the University of Edinburgh, being shown the way by local fossil expert, Dugald Ross. A crofter and builder, Ross grew up on Trotternish. As we found ourselves drawing level on the slippery mud path, I greeted him in Gaelic, his native tongue and one I know shamefully little of for a Highlander. In sparse and gentle island lullabies he began to share the potted history of his life.

As a teenager, Ross had made many archaeological discoveries among the forgotten settlements on the Island. Arrow-heads, once believed to be carved by faeries, pointed to the most ancient settlers of

---

[*] A relative term. To Highlanders, Skye's prints are much less remote than those on the south coast of England.

the Inner Hebrides. Inspired by these finds and the rich local heritage, Ross had claimed the ruined walls of an abandoned one-room schoolhouse near his home, and declared to his father – who was a stonemason – that he would restore it and build the island's first museum. Ross's father scratched his chin and shook his head, leaving his son to pursue the wild dreams of a young man.

A lifetime later, Dugald Ross's museum is still standing. It is a stone-walled, single-storey construction, perched by the roadside. Under the slate roof the insides are whitewashed stone and the air leaves a tang of rust on the tip of your tongue. It is filled with the lovingly curated debris of Skye's history. Farm equipment saved from roadside ditches or pulled from the peat. Glass bottles, floral crockery and measuring tools, a rusted mole trap. And fossils big and small.

Since discovering his archaeological treasures, Ross has made many significant Jurassic reptile discoveries on the Isle of Skye. These included fragments of dinosaurs and their cousins, the marine reptiles. He has been published many times along with his friend and collaborator, Neil Clark, curator of the Hunterian Museum in Glasgow. Between them over the last 20 years Ross and Clark have described individual bones and footprints exposed along the magnificently ruinous shoreline of northern Skye. Latterly, Ross began working with researchers at the University of Edinburgh too, and this was how I met him that sunny spring day.

As we walked towards Rubha nam Brathairean, Ross told me it meant 'Point of the Brothers'. There were two stories behind the name, he said: either it was named for a monk who lived there, or it was named for the tragic death of a local fisherman's brother. 'So which one is it?' I asked him. Like many place-names it's hard to separate its literal origins from the poetic. With a whiskery smile Ross invited me to pick whichever I liked best.[*]

---

[*] Through some impressive detective work, Ross has since discovered that Rubha nam Brathairean is probably a corruption of Rubha na bràthan (Quern-stone Point). Querns are large circular stones once commonly used for grinding grain, and the sizeable man-made circular holes in the rock suggest it was once the site of industrial-scale quern-stone removal.

Amateur fossil hunters in Great Britain are well acquainted with the fossil-rich sites of the English coast. In a trend initiated by Victorian naturalists, they walk the shores in search of ammonites, belemnites[*] and the occasional marine reptile or dinosaur that emerges from the salt-rubbed shores. Here in Scotland, as one might expect, our Jurassic coast is a lot less civilised. You can't amble, it's not a gentle beach stroll. Skye's northern coast drops like a ledge at the end of the Earth into the salty Minch, the cold seaway running between the Hebrides and the mainland. At the base of these cliffs the water batters at the sandstone, tearing off chunks of it in winter fury and tossing them along the shore like cabers.

The sun was shining and the sea quite calm as our team hiked to the shore to search among the geological wreckage. This was very different from my work with the main team in the south of Skye. Here on Trotternish we clambered over the massive sandstone boulders like ants on sugar cubes. Vertebrate fossils are very rare in northern Skye, and most days we came back with nothing but scraped knees and raw, wind-whipped cheeks. Along much of the coast there aren't even the ammonite riches that compensate avid fossil hunters on English shores for their diligent hours.

Ross has spent years negotiating these coastlines. Although his finds are significant for the Scottish fossil record – and have rightly earned him the Palaeontological Association's Mary Anning Award in recognition of his achievements – numerically they are small compared with the long-established sites in England. But in the same way centuries of crofters have laboured hard to eke a livelihood from the harsh coastal landscape, so the same tough labour searching for fossils here is worth the reward. It is home, which makes every find a treasure.

The geology of Skye is unique. Most of the island is smothered in a blanket of volcanic rock – bad news for palaeontologists, because it

---

[*] Ammonites are the spiral fossils of marine invertebrates. A soft-bodied animal lived inside them, similar to the nautilus; it would have looked like an octopus living in a snail shell. Belemnites were squid-like animals, but what usually remains of their bodies in the fossil record is the bullet-shaped hard internal skeleton.

almost never contains fossils. The rocks we want are underneath this blanket, poking their toes out into the sea around the edges of the island. They are sedimentary, having been laid down when waterways dumped their loads of sand, grit and clay, scoured from the high lands upstream. Most of the layers are Middle Jurassic, from a time called the Bathonian, around 166 million years ago – although some slightly younger and older pockets of rock survive scattered here and there like lost pennies. Taken together, these sedimentary rocks and the volcanic lid that tops them are not just visually stunning, but also tell breath-taking story of millions of years of landscape formation, destruction and development.

Looking at a map of our planet as it was in the Middle Jurassic, the pattern of continents hints at the layout we know today. The British Isles were part of a chain of islands sitting at the same latitude as the Mediterranean does today, and surrounded by a seaway rich with life. These waters connected the warm tropical Tethys Sea in the south-east to the cold Boreal Ocean in the north. Like similar marine habitats today, this mixing zone of hot and cold water would have stirred up a soup of nutrients and overflowed with life. These rich marine ecosystems are preserved in the fossil record across the once-waterlogged European continent.

These Jurassic islands were lush with ferns, cycads and dense stands of horsetails. It was warmer than today, semi-tropical and humid. On higher ground, conifer forests coated the hillsides. There were no grasses or flowers, because neither had evolved yet, but the cycads and a similar-looking group of extinct plants called bennettitales produced a kind of fruiting body that would have attracted swarms of scorpion flies and lacewings, and in turn, small vertebrate animals like mammals and lizards, which fed on them.

Similar to today, Skye at this time was sandwiched between the high hills of mainland Scotland to the east, and the backbone of the Outer Hebrides in the west. Over the course of the Middle Jurassic Skye rose and fell as the tectonic plates shifted restlessly beneath. Sometimes the landscape was exposed, producing rich terrestrial habitats interspersed with lagoons. At other times the sea swallowed it again. As a result, the rocks alternate in their character and contents. There are layers of footprints captured in exposed coastal flats, along with the bones of dinosaurs, pterosaurs, and smaller reptiles like crocodiles and turtles, alongside mouse-sized mammals and salamanders. Other layers are peppered with spiral-shelled ammonites, belemnite bullets, and fish and marine reptiles including plesiosaurs and ichthyosaurs: the contents of a plentiful shallow sea.

Throughout the Mesozoic – the 150 million years that comprise the Triassic, Jurassic and Cretaceous time periods – rocks continued to form and erode via the usual mechanisms of geology. At the end of the Cretaceous, 66 million years ago, the asteroid-induced mass extinction that wiped out incredible reptile groups such as non-bird dinosaurs (as well as ammonites and other staples of the Earth's ecosystems) did the same job here in Scotland. The survivors, including mammals and birds, repopulated the landscape of the new era: the Palaeogene period.

For the ecosystems of the Hebrides however, another disaster followed hot on that space-rock's heels. Around 55 million years ago, Europe and North America were in the middle of a messy tectonic divorce. Having lain together for half a billion years,[*] they began to drift apart. Along the edges of the shifting plates, volcanoes wept. The remains of these magma chambers can be seen on the Scottish west coast in places like the Ardnamurchan peninsula – essentially the guts of a giant volcano sticking out into the sea. The Cuillin mountains are another remnant. The volcanoes spewed their contents over the Inner Hebrides and smothered the Mesozoic sediments and the fossils they contain in dark, hard basalt.

Reaching the shore on Trotternish, our team began searching. We already knew there were footprints, discovered the previous year and ready for further study. I'd never seen dinosaur footprints 'in the wild' before. Would I know what to look for? At first they resembled random rock pools, but quickly the pattern became obvious: left foot, right foot, a rhythmic impression across the rock platform. My heart picked up pace. It was similar to the experience of reaching a muddy patch on a woodland path and seeing a paw-print impressed in the earth – something went this way before you. In this case, that

---

[*] I'm sure many marriages feel that long.

thing was a dinosaur, and before was over 166 million years into Scotland's past. I followed the prints, placing my feet into them. Football-sized round holes, smaller triangular dents. Striding here, pausing there, front foot with a hind placed hot on its heels. Stage directions for a ghost.

Pausing in reverie and looking north, I admired the waterfall that throws itself over the precipitous drop of Kilt rock from the flat fields above. These dramatic landscapes are among the most iconic in the UK. The Old Man of Storr – a vertical finger of basalt pointing skywards from the hillside above the road – is a visual synonym for the magical fingerprint Scotland leaves on so many visitors' hearts (it also appears in lots of film and television, including Ridley Scott's alien blockbuster, *Prometheus*). I was standing in two Scotlands at once: Skye in the harsh Anthropocene, where my colleagues were pulling up their hoods as the first of many rainclouds emptied on our heads; and Skye in the warm Bathonian, where a herd of herbivorous sauropods made their way across a brackish lagoon in the Jurassic sunshine, heading for rich pickings on a nearby islet.

Thankfully, NatureScot have placed strong protections on Skye, recognising the geological uniqueness of the island through its system of SSSIs, including my team's site in the south. In 2019 the Scottish government signed a Nature Conservation Order for Skye, providing enhanced legal protection against unofficial collection of the fossils. This helps deter those who would steal our nation's heritage for profit, damaging the landscape in the process.

Research into patterns of animal evolution in the Mesozoic suggests that many of the main groups of animals on our planet today can trace their direct ancestors to the Middle Jurassic. There appears to have been an explosion in animal diversity. This pattern holds for dinosaurs – the famous long-necked sauropods and bipedal theropods, for example, and the ancestors of birds can be traced back to branches that fork at this time. The same is true for mammals, as well as marine and flying reptiles. Looking at the diagrams of animal ancestry, it is as though someone planted a bomb in the centre of the Jurassic, and its explosive dust cloud of diversity spread out through the rest of the Mesozoic.

On Trotternish, fossils are not that common and to be honest, most are not very impressive to look at. But their age makes them important. As well as coming from a key time in evolutionary history, Middle Jurassic fossils are globally rare. This means that although Skye's finds are superficially less appealing,[*] they provide significant scientific information for researchers. All you need is the technology and expertise to uncover their message.

By the end of the day on Skye the team had found several scraps of interest, and even stumbled on a new set of footprints. Slowly, we were putting together a picture of prehistoric Scotland from these scattered fragments. Each one helps us see the Middle Jurassic more clearly, to understand what drove this explosion in the diversity of life on Earth.

Although dinosaur feet stride easily into the headlines, these animals provide only a partial picture of the ecosystem they inhabited. Imagine trying to understand the grasslands of the Serengeti, but only studying the lions and wildebeest. Or looking at a tropical reef and watching nothing but the sharks. As wonderful as these creatures are, to truly understand an ecosystem and how it functions, we have to investigate the smallest inhabitants that often play key roles.

Like most palaeontologists, I began my career thinking that the most exciting extinct creatures were massive Ice Age mega-beasts and colossal Cretaceous reptiles. I soon realised that the real meat and bones of the past are to be found in the tiniest of lifeforms.

Mesozoic mammals are one of those pivotal groups. Despite their generally diminutive size during the time of dinosaurs, their evolutionary path is not one of relegation, but *innovation*. They seized the moment with endothermic gusto. What's more, as I began my own path into science I saw that their small size during one phase of prehistory was only a snapshot in a much longer, more complicated story. They were key players in the birth of modern ecosystems, but

---

[*] There are some stunning exceptions, many of them currently on the cusp of publication, including mammals, pterosaurs, and lizards.

also first-past-the-post for novel modes of life long before, modes such as dedicated herbivory and hyper-carnivory. Their family album was massive and filled with unlikely cousins from a long-lost world.

This branch of the tree of life has seen some of the most radical changes of all. Their tale is packed with evolutionary eurekas that would, one day, produce all the mixed blessings of humankind.

CHAPTER TWO

A Thoroughly Modern Platypus

'That is the beauty of it all,' said the Kangaroo, 'the Platypus is so learned and so instructive, that no one tries to understand it; it is not expected that anyone should.'

Ethel C. Pedley, *Dot and the Kangaroo*

It was a cold February day in London in 1824. Wrapped in a long black cloak with his scarf tucked tight against the chill, Reverend William Buckland entered 20 Bedford Street. His top hat hid a receding hairline, although he was only 40 years old. Beneath the rim he regarded the entrance hall with heavy-lidded chestnut eyes. His plain features belied his exceptional eccentricity: it was a wonder he hadn't ridden into the building on horseback, as he was known to do in his dramatic lectures at the University of Oxford. But today was a serious affair. Tucked under his arm were papers in preparation for his first address as the new president of the Geological Society of London.

The Reverend was about to announce the world's first dinosaur.

Buckland's *Notice on the* Megalosaurus *or the Great Fossil Lizard of Stonesfield* was a key moment in palaeontological history. 'I am induced to lay before the Geological Society ... parts of the skeleton of an enormous fossil animal,' he said. 'The beast in question would have equalled in height our largest elephants, and in length fallen but little short of the largest whales.'[1]

Geologists at the time knew that the Stonesfield Slate, the rock that had produced these bones, pre-dated humankind. They were from a time 'before the flood'. But there was no cohesive theory of evolution or acceptance of extinction to explain their existence. The discovery of enormous bones in ancient rocks was world-changing. In the following years more fossils were recognised as belonging to similar huge reptiles. *Megalosaurus* would be a founding member of what the anatomist Sir Richard Owen later called the 'terrible lizards', *Dinosauria*.

Fossil mania swept the Victorian world, and the reptiles were especially compelling. Sculptures of the brutes were commissioned to skulk in the grounds of the Crystal Palace in London, site of the World Exhibition in 1851. To mark the event, a dinner was held in the open body of one of the sculptures, and guests were sent their invitations on artificial 'pterodactyl' wings.[*] By the end of the nineteenth century the very public squabbles between American palaeontologists hunting for new specimens had excavated a permanent place in Western culture for dinosaurs and the men who fought over and described them (there were women too, but their contributions have been under-represented until recently; we'll return to that subject later).

But there was something else to be found in the Stonesfield Slate, a far subtler beast. Buckland's announcement included a second animal, one substantially more modest than its gigantic 'lizard' cohabitant. It wasn't as long as a whale, and couldn't be climbed inside for a dinner party. Although it failed to capture the public imagination in the same way, among scientists it was 'not less extraordinary' than the giant reptile – in fact it was so surprising and controversial that Buckland admitted he 'should have hesitated to announce such a fact, as it forms a case hitherto unique in the discoveries of geology'.[2]

This startling fossil was the jaw of a mammal.

One scorching summer day almost 200 years later, I held that fossil jaw in my hands. It is now in the collections of the Oxford University Museum of Natural History. I was sitting at a scratched workbench in the shadow of towering dark wooden cabinets. A faint whiff of preservative pervaded the cool room, like nail varnish remover with the top left off. Beside me, a large window overlooked the crowds of visitors outside. They entered and left the building like honeybees at a hive. Even today we still flock to gawk at the bones of dinosaurs and other extinct animals, and the Oxford Museum punches well above its weight in fossils. Hundreds of dinosaur bones have been collected

---

[*] Pterodactyls are a type of pterosaur, not dinosaurs, but flying reptiles closely related to the dinosaurs.

since Buckland's day, and for adults and children alike seeing their enormous skeletons is an exciting and moving experience.

One storey above them, I was having my own emotive moment in the fossil mammal collections. The jaw that Buckland had been shy to reveal is only about two centimetres in length. The bone is a slightly darker, cappuccino brown than the surrounding rock, which is fine-grained and dust-coloured, peppered with little white dots. It felt grainy in my hot palm. Around the jaw the rock has been carefully scraped away by museum preparators, revealing the edges of the bone sunk into the stone. The teeth are tiny, like couscous grains. They top a deep-set jaw. Some of the bone has been removed to reveal the tooth roots, reaching down to anchor themselves against a lifetime of biting.

A typical small mammal jaw. It was like any of today's hedgehogs or opossums, perfect for grabbing and crunching up insects. What could possibly be controversial about that? The problem wasn't the fossil itself, but the rock it was encased in. It was marine, and far too old to find mammal bones in. How on earth had it got there?

The Stonesfield Slate in Oxfordshire in which *Megalosaurus* and the mammal jaw were found had been mined for roofing tiles since at least 1640. The rock was dug up in the summer and laid out on the ground through the winter, where it was wetted and left to split along the seams by the growing crystals of frost.[3] Alongside building materials, the rocks of Stonesfield were producing the bones of ancient animals.

There were several mammal jaws found before the one noted by Buckland. The first jaw was actually discovered around 1764, but the owner had no idea of its significance, probably assuming it belonged to a small reptile. This jaw was passed on to several owners before ending up in the museum of the Yorkshire Philosophical Society, where it was rediscovered much later. Meanwhile, more jaws were uncovered, but it wasn't until one of them was passed to Buckland by a student that anyone turned their attention to these startling oddities.

Buckland didn't shirk from oddity – he was an oddity himself. He had travelled extensively on his geologising horse across the country, and was often clad in academic gowns while at the rock face. Born to a Church of England rector in a parish in Devon, Buckland had been taken to local quarries by his father as a child. There he was immersed in the world of fossils, especially the spiral shells of ammonites.[4] Although pursuing divinity, Buckland had always been drawn to study the natural world. Throughout his life he would surprise people with his eccentric mannerisms and outlandish home full of animals, both stuffed and alive. He once licked a church floor to confirm that a puddle on it was, indeed, bat urine. Yet even an open mind like Buckland's struggled to make sense of these incongruous mammal jaws.

Although geology was still in its infancy, key mechanisms had been identified which would be crucial to the growing understanding of how fossils fit together to tell the story of life. In Scotland by the late 1700s, the visionary geologist James Hutton had transformed common conception of geological time by identifying the achingly slow processes that formed, eroded and reformed rocks. Around the same time a mining surveyor in England named William Smith had begun to notice that rocks lay in recognisable sequences of layers, and that the presence of certain fossils could be used to distinguish between otherwise similar-looking strata. He used this principle to match rock sequences across the country, producing the first geological maps, including one of England in 1815.

While Smith was mapping England, French researchers had arrived at the same conclusions during their work on the formations around Paris. The concept that fossils can be used to correlate and date rocks is what we now call biostratigraphy, and it remains a vital component of geology and palaeontology. With this new realisation, the race to record the stratigraphy of both the known and the 'newly discovered'* world began.

The Stonesfield Slate of England lay among the Oolites, or egg-stones, rocks that got their name from a textural resemblance to fish eggs. The Oolites belonged to what was then known as the Mid-Secondary age, referring to an old system of geology that recognised the oldest Primary rocks (what we now call the Palaeozoic), then the Secondary ones above them (the Mesozoic), the Tertiary rocks on top of that (most of the Cenozoic, namely the Palaeogene

---

* That is, new to Europeans, not to the peoples who had lived there for millennia.

and Neogene), and finally the youngest Quaternary layers like icing on the surface.

Buckland, like many of his contemporaries, believed the Earth in the Secondary age was a predominantly marine, water-logged environment. Ichthyosaurs and plesiosaurs, marine reptiles that are only distantly related to dinosaurs, were already well known – many of them found by the astute and prolific fossil collector, Mary Anning. Anning not only found the fossils, she knew as much about them as the men who bought them from her and described them. Until recently her role was downplayed in the history of palaeontology, but she has now received recognition for her massive contribution to science.

Anning's reptile fossils and the copious aquatic invertebrates she also found, such as ammonites, proved that Secondary rocks were marine. The name given to these rocks, Jurassic, was chosen by the naturalist and explorer Alexander von Humboldt in 1795, in reference to the limestone rocks of the Jura Mountains on the Swiss-French border. Limestone, by its very nature, is a rock formed underwater: it is produced on the sea floor by the compacted snowfall of tiny marine organisms like coral, molluscs and forams.

How could a tiny furry mammal have lived in this ancient water-world? Had there been a mistake? Buckland was so uncertain that he sought the opinion of France's foremost anatomist and savant, Georges Cuvier.

Pictured in the frontispiece of his books, Cuvier's pouting lips, pale wavy hair and the elaborate ceremonial jacket of the Académie des Sciences give the impression he was a dashing Napoleonic dandy. He was in fact one of the most influential scientists[*] in the Western world, and credited as the 'Father of Palaeontology'.

Tutored by his mother, Cuvier had excelled academically since childhood, devouring the major works of zoology and biology so that by the age of just 12 he was already an expert naturalist. After graduation his talents were soon recognised, and he was invited to Paris. There, he rapidly distinguished himself through his studies of fossil elephants and other extinct creatures.

---

[*] The term 'scientist' was not coined until the 1830s, and not widely used until the twentieth century. I use it here for the understanding of the modern reader.

What set Cuvier apart was both his insurmountable knowledge, and his use of comparative methods to understand organisms. He realised that not only could bones be compared to distinguish different species, but their shape was often similar too, if they had the same lifestyle. Digging animals, for example, had similar skeletons even if they were not close relatives, with broad fingers and big arm bones to accommodate lots of muscle attachments. In other words, their *function* was linked to their *form*. Using these methods, Cuvier could famously work out from the scantiest amount of skeleton both what an animal was and how it had lived. A dazzling party trick, but as we will see in a moment, he was not always correct.

Cuvier was a powerful personality and a considerable scientific authority, but more importantly he was eloquent and accessible. In the early 1800s, he published a series of comprehensive works that drew together the latest understanding of the history of life and how it related to geology. He presented the evidence for Earth's changing environments as seen in the rocks, and placed it alongside his unparalleled knowledge of the fossil record. But above all, he presented his works in a way that was succinct and readable to the well-educated Western public, as well as to other scientists. This was one of the many reasons that he became internationally renowned, and his ideas dominated the biological and zoological sciences during the first half of the century.

In 1818 Cuvier was visiting England from his home country to view the growing museum collections. He met Buckland in Oxford and proceeded to examine the palaeontological collections there. When it came to the new mammal jaws from Stonesfield, Cuvier identified them as belonging to a marsupial opossum, based on similarities in the shape of the jaw. Buckland therefore announced the mammal jaw from Stonesfield as *Didelphis,* the North American opossum's scientific name. This left contemporary scientists to puzzle how a marsupial mammal, of the kind now only found in Australia, Indonesia and parts of the Americas, had found its way into the reptilian Jurassic rocks of England.

For Cuvier and his followers however, a mammal jaw of any kind could *never* be found in Secondary rocks. It seemed clear from the fossils known so far that there had been an 'age of reptiles', and then later, an 'age of mammals', one distinctly separate from the other. For one of Cuvier's students, the answer to the Oolitic mammal conundrum was simple: the Stonesfield Slate had to be Tertiary, not Secondary after all. He suggested that the English stratigraphers had made a mistake, and misidentified a younger pocket of rock.

As you might imagine, this opinion wasn't popular with the gentleman geologists of Britain. But whatever the truth about these inconvenient little mammals, it was a mere blip in the grand theory of the 'age of reptiles' and was, for a time, easily ignored.

Cuvier was a 'catastrophist'. Catastrophists believed the Earth was periodically wiped clean by catastrophic events such as floods, which caused the extinction of all life. In this narrative the Secondary rocks recorded the age of reptiles, which was followed by a catastrophe that led to the Tertiary age of mammals, and latterly a flood and the age of man. According to this world-view, no mammal could have existed alongside the giant reptiles. With each catastrophic cleaning of the slate a new fauna took over the Earth – but certainly *not* by evolution.

Cuvier famously didn't believe in evolution, an idea which at the time still took the form of 'transmutation'. Jean-Baptiste Lamarck – a contemporary scientific heavyweight – argued that transmutation was driven by a constant increase in the complexity of life forms, ultimately leading to perfection.* But naturalists like Cuvier thought that animals and plants fitted perfectly into their way of life already, and had no need to improve: form and function were inextricably tied together. If creatures were so intimately tied to their position in nature it seemed impossible that they could change, otherwise they would be unable to survive.

Another of Cuvier's main objections to transmutation was a lack of what we would now call transitional fossils. Most of these so-called 'missing links' would eventually be found over the next century and

---

* Lamarck is famously lambasted for getting evolution 'wrong', but his ideas weren't as incorrect as is often portrayed. He argued for mechanisms of 'use' and 'disuse', believing those characteristics that were used would be passed on, and those that weren't would atrophy, which isn't all that far off the mark.

beyond, proving incontrovertibly that animals did indeed change over time (which Darwin and Wallace later explained with their theory of evolution).* But in the early nineteenth century there were still too many gaps for people like Cuvier to be convinced. Scientists and the public alike found Cuvier's link between form and function – proving an animal's perfectly designed suitability for its lifestyle – the more persuasive argument explaining the patterns of life on Earth.

A student and colleague of Cuvier's, Étienne Geoffroy Saint-Hilaire, disagreed with his tutor and favoured transmutation. His explanation in support of it was intimately linked to the story told by geology. Saint-Hilaire argued that transmutation was not tied to a vague drive towards perfection, but to an organic and natural response to environmental change. Such change could be seen in the rock record, and provided a mechanism that laid the foundation for Darwin's thinking decades later.

Unfortunately, to test his ideas Saint-Hilaire performed Frankenstinian experiments. He used chicken embryos, manipulating them in the egg to produce 'monstrosities'. While this may have lent some support to the idea that changes in surroundings could cause changes in living things, it did nothing for his street cred – the Victorian public found his experiments disturbing. What was more, his argument seemed to suggest that rather than obeying a guiding divine hand, life lurched along thanks to a series of unsavoury accidents. It was an unappealing notion in a society that believed in the betterment of humanity and nature through industry and innovation.

Running beneath the vehement arguments against transmutation was a deep vein of anxiety over human origins. It was utterly repugnant to the Georgian and Victorian mindset to contemplate that humans might be related in some way to other animals. Not only was it offensive to humankind, it was offensive to God and his glorious

---

* The concept of 'missing links' is based on the nineteenth-century idea of the great chain of being. We now know life evolves in a network of strands that diverge, and this term is defunct. Like 'living fossil', it is not a meaningful term, and actually quite misleading.

creation. Humankind – at least the white, Western part of it – was made in God's image.* For mainland Europeans like Cuvier and Saint-Hilaire, the political and social revolutions of the 1700s had separated Church and State and allowed scientists greater freedom from biblical doctrine. But in Great Britain, finding ways to integrate the emerging understanding of science and nature in a Christian framework remained a persistent preoccupation.

Growing up in this atmosphere of reconciliation between science and established faith was a young Scot who would challenge Cuvier, transform our understanding of geology, and latterly embrace the evolutionary theory of his good friends Charles Darwin and Alfred Russel Wallace. His name was Charles Lyell.

Lyell came from a wealthy family from Forfarshire near Dundee, and like other rich white men he pursued leisurely interests like literature, botany and the study of rocks. He attended Buckland's lectures during his time at the University of Oxford, and eventually gave up his career in law to become a geologist – after all, his wealth meant he could do anything he pleased. In paintings and photographs, Lyell's massive forehead looks fit to burst with all the knowledge being absorbed and combined inside it. Like Cuvier, he was a gifted communicator, but was also brilliant at bringing together the ideas of many great minds and synthesising them to see the bigger picture.

In the 1820s and 1830s, Lyell became an increasingly integral part of the scientific scene both at home and abroad. He was among the first who realised the Earth had not been an ancient water-world in the past, as previously suggested. Travelling extensively across Europe, Lyell soaked up the geology of the continent. He laboured and argued with those working on geological problems in France, Italy and Austria, and spent many days combing through museum collections examining invertebrates and minerals. He climbed Mount Etna, and logged the stratigraphy of high mountains and deep ravines. 'The whole tour has been rich ... in those analogies between existing Nature and the effects of causes in remote eras,' he wrote in a letter to his father. Lyell could see the connection between the processes

---

* We will see in later chapters how the search for human origins led to the perpetration of horrific acts in the name of science.

occurring in the present, and the rock record of the deep past, and that 'will be the great object of my work to point out'.[5]

Lyell's book, *Principles of Geology*, was written as a result of these travels, and its three volumes were first published between 1830 and 1833. Its driving theme was that the Earth had been shaped over immense stretches of geological time by the same slow-moving processes that were visible today: the erosion of hillsides, the movement of soil by rivers, the steady building up of sandbanks. This was a stark contrast to Cuvier's abrupt catastrophism. Lyell insisted that the reason it seemed like the Earth had been exceptionally watery in the past was simply a bias in the rocks: marine environments were more commonly and easily preserved than terrestrial ones. He was right.

But like his contemporaries, Lyell wanted to protect humankind from the distasteful ideas of transmutation. His approach was to use geology to show that life was not directional at all. He began by questioning the evidence that Cuvier and others had presented from the fossil record. He thought fossils were too sporadic and unreliable to support this grand story of progression. The mammal jaws of Stonesfield, far from being a mere anomalous oddity, seemed to him a clear sign that Lamarck's ideas about transmutation were also faulty. If transmutation were true, surely marsupials wouldn't remain unchanged over such vast timescales?

While Lyell was speculating about all this, his friend William Broderip found and published details of a second mammal jaw from Stonesfield – a different species from the first. An international team of geologists took a second look at the stratigraphy of Stonesfield Quarry and agreed that it was indeed of Secondary age. The mammal jaws were Jurassic after all and there had been not one, but *two* types of opossums living in the Oolite. This was a great comfort to Lyell. 'So much ... for the theories of a gradual progress to perfection!' he wrote to his father in 1827, 'there was everything but Man even as far back as the Oolite'.[6] In other words, the horrible implications of transmutation couldn't be correct because there were marsupials, unchanged, in the time of reptiles.

In the eighteenth century the idea of extinction had seemed inconceivable and disturbing. Why would life be created by God, only to be snuffed out again? Many Western thinkers argued that the animals whose bones were being found in ancient rocks must still be alive elsewhere on Earth – we just hadn't found them yet. Lamarck also couldn't believe that animals could completely disappear, which was why he argued they must have changed rapidly into other forms. The occasional discovery of 'living fossils' seemed to justify the view that everything was still here, *somewhere*.

But as Europeans invaded and explored the rest of the world it became clear that there were no hidden valleys full of mammoths,* or undiscovered ammonite colonies hiding in distant reefs. More and more bones were turning up from creatures that clearly no longer existed.

Cuvier may have pushed hard against transmutation, but he was the first to put his neck on the line and argue that life could become extinct. By pulling together the compelling new evidence from the fossil record, geologists and palaeontologists like him persuaded the world of the reality of extinction.

Convinced by all of this evidence, Lyell argued in the second volume of *Principles* that species arose and sometimes became extinct, and that this could be seen in biostratigraphy, as organisms appeared and disappeared through the layers. As to the origins of these species ... he hedged his bets and left it ambiguous.

If there's one thing you can say about nineteenth-century scientists it's that they loved big theories to explain everything. From Lamarck to Cuvier to Lyell, people were searching for all-encompassing rules governing the world they observed around them. Buckland was no different. One of his greatest contributions to this century of big ideas was the direct result of the mammal jaws of Stonesfield.

Now that they were proven to be ancient in origin, and to have lived in a world that had both terrestrial and marine environments, the Stonesfield mammals' appearance in the 'age of reptiles' still required an explanation. Buckland would find this in the writing of an intense young British anatomist, whose star swiftly rose to eclipse Cuvier's after the Frenchman's death in 1832.[†]

---

* The first identification of American mastodon – which were similar to mammoths – remains as belonging to a type of elephant was made by enslaved Africans in the early 1700s. Many of them came from Angola or the Congo, where elephants were well known to them.

[†] Aged just 62.

Richard Owen was born in 1804, the son of a rich merchant from Lancaster. Trained as a surgeon, he found himself working instead as a museum assistant. With a sharp mind and dedication to comparative anatomy, he was quickly catapulted through the ranks and spent his whole career pursuing the scientific understanding of animal anatomy – including fossils. The Natural History Museum in London ultimately owes its existence to him. This was the man who would coin the term *dinosauria* in 1842.

But Owen had a dark, unpleasant streak through his character. He single-mindedly demolished those he disliked or disagreed with. Later in life, he became a cantankerous and bitter person, struggling to accept that the scientific world had moved on without him. Nevertheless, his contributions to science still underpin our understanding of mammal evolution.

Mammalia, the taxonomic group that includes humans and other milk-producing animals, was named in 1757. The term literally means 'of the breast'. Although mammals also possess hair[*] and have a unique jaw and ear anatomy which we will learn more about later, it was the suckling that Carl Linnaeus, who firmly established the science of taxonomy, chose to emphasise. Of the six major groups named by Linnaeus in the 1700s – which included *Aves* (birds) and *Amphibia* (amphibians) – it is the only one named for a feature possessed by only one sex, making it a particularly political work of taxonomy.[†]

Mammals today are split into three main groups: placental mammals, marsupials and monotremes. We humans belong to the placentals, and broadly speaking they are distinguished by retaining their growing embryo inside the womb, nourished by the placenta. Marsupials also have a placenta while in the womb, but their offspring are born much earlier and usually complete their development in a

---

[*] While it's true that some mammals have lost most of their hair, even whales retain an echo of its presence. Baby whales often have a few whiskers on their chins, which fall out as they grow up.

[†] There is a social aspect to the choice of name. In the 1700s most wealthy women employed a wet-nurse to breast-feed their children, but in Linnaeus's lifetime there was a drive to encourage women to breast-feed their own children. This cultural backdrop undoubtedly influenced his choice of name for the group, placing emphasis on something that was being pushed as the 'natural order'.

pouch (the name comes from the Latin for pouch, *marsupium*). Most mammals alive today are placentals, and almost all of the rest are marsupials. We'll return to the monotremes – the only egg-laying mammals, including the platypus and echidna from Australasia – in a moment.

In 1834 Richard Owen described in exceptional detail the reproductive organs of the kangaroo. He compared them with other marsupials, placental mammals, monotremes, and finally with reptiles and birds. Marsupial organs, said Owen, resembled monotremes more than other mammals. According to his observations, monotremes and marsupials together bore some resemblance to reptiles in their reproductive organs.* Coupling this with other peculiarities of their anatomy, he then asserted that 'it is in *Mammalia* that the brain is perfected: we can trace through the different orders the increasing complication of this organ, until we find it in man to have attained that condition which so eminently distinguishes him from the rest of the class.'[7]

Buckland, reading Owen's impressive detailed comparisons, agreed with his conclusion that marsupials represented an 'intermediate place' between reptiles and placental mammals. Monotremes, which were considered even more inferior than marsupials, must inhabit an even lower and earlier position in the history of life. This seemed to explain how a marsupial mammal could appear in the age of reptiles: they had inhabited a world that was still too primeval for the advanced placental mammals. The Stonesfield marsupials were the primitive forerunners of the grand 'age of mammals'.

When Buckland reconstructed the animals of the Secondary rocks in 1836, drawing two neat little marsupial mice crouching beside giant flying and marine reptiles, the picture made sense to the public and scientists alike. It was logical to assume that marsupials existed alongside dinosaurs in a primordial version of the world, ultimately to

---

* While the idea persists that monotremes and marsupials are more like reptiles than placental mammals, to the modern anatomist there is little similarity. They might possess a cloaca, but it isn't structured the same way at all. As you'll find out as you read on, monotremes and marsupials are no more reptilian than you or me.

be replaced almost entirely by the superior placental mammals. This idea of 'primitive' versus 'advanced' mammals, with life becoming increasingly superior over time, was arguably one of Buckland's most persistent contributions to our view of life on Earth.

This picture was based on many misconceptions, as well as the imperialist outlook of European empire. A key flaw was the notion that marsupials and monotremes were somehow less advanced than other mammals. It's an idea so entrenched that it persists into the twenty-first century. But it's just not true.

The history of the platypus is a tale of slaughter and misunderstanding.[*] The first platypus was sent back to England from Australia in 1798. Because the naturalists and anatomists who first described these new species were white Europeans they were inherently biased in their observations of the animals sent back to them from the New World, and dismissive of local knowledge about them. They were, however, reliant on local expertise to find and capture animals (as well as locating and digging up fossils). Despite this, indigenous knowledge was rarely credited afterwards.

With no monotremes or marsupials in the Old World, placentals were considered the norm and anything deviating from them was aberrant. Scientific as well as public opinions and descriptions emphasised Australian animals as weird and bizarre – a trope that continues.

Having said that, there is no doubt that the platypus, *Ornithorhynchus anatinus*, is unique in the modern world. Along with the echidnas, *Tachyglossus* and *Zaglossus*, these mammals are now the only ones that lay eggs. This fact took considerable time to be proven beyond doubt to the sceptical European scientific establishment, and was taken as evidence in support of monotreme 'primitiveness'. Platypus males and some females have venomous spurs on their ankles, and although they produce milk – the hallmark of all mammals – they are nipple-less, instead extruding it through patches of the skin, to be licked by their offspring. For years it was debated whether the platypus and echidna could produce milk at all.

---

[*] For a full account of it, I recommend *Platypus*, by Ann Moyal.

The group name, Monotremata, means 'single-hole', referring to their lone entry and exit point for excreting, mating and laying eggs. It's an anatomical set-up more commonly associated with birds and reptiles than mammals. In some ways it's understandable why people thought of them as antiquated or even 'reptilian' for retaining these characteristics, some of which have been handed down from their mammal ancestors in the Early Mesozoic. People have even used the term 'living fossil'.

But there's no such thing as a living fossil. Monotremes are just as 'advanced' as the rest of us. Natural selection doesn't stand still: genes amble on, modifying and recombining with each generation. Although some animal groups may superficially appear little changed from their ancestors, in every case closer inspection reveals a host of differences. The idea of 'primitive' or 'advanced' doesn't exist in the biological sense, there is only change over time, and evolution also has no ending point, so life neither improves nor degenerates.

Among the random variations that appear naturally in a population of animals, those that aid them to survive at any given time are the best adaptations for that population, in that environment, at that moment. In some populations this eventually leads to large changes with visible and radical alterations in physiology and anatomy. In other groups the changes are subtler, so they appear like relics from the past to an outside observer. But inside, the cogs of biology endlessly turn. You can be sure that if what they are doing works, it is just as honed and 'advanced' as any other trait or group persisting on the Earth today.

In a pre-Darwinian world, the mechanisms of natural selection were unknown to those scientists who first examined monotremes and marsupials. While Owen was describing kangaroo privates in his laboratory, Darwin was still on board the HMS *Beagle*. His round-the-world journey as a gentleman naturalist on this Royal Navy sloop would provide the experiences and data he needed to understand the processes of evolution. With this knowledge, and a century of research to enhance it, we can now see monotremes for the modern wonders they are.

Monotremes do retain a few characteristics from earlier mammals that have since been altered in other mammal groups. Despite this, they are what scientists call 'highly derived', meaning that they are a long way from the base of their family tree. The famous 'duck-bill' snout of the platypus for example, is a completely unique food detection device that has not evolved previously in the entire history of mammal kind.*

The semi-aquatic platypus's scientific name means 'bird-like duck-snout', but it's a misnomer. Unlike a bird's beak the platypus snout is soft and bendy. What no one knew when they named it is that the entire snout is stippled with over 40,000 dot-like mechano- and electroreceptors, a Milky Way of sensitivity. The mechanoreceptors provide a unique sense of touch. As the platypus dives beneath the freshwater rivers and lakes of mainland Australia and Tasmania, it closes its small eyes and lobeless ears (located just behind the eyes), and the nose takes over. Like a hand reaching out in the dark, the platypus uses its snout to feel for food along the river bottom. Madly darting that tactile snoot between the rocks, the platypus employs a second line of advanced hidden technology: rows of electroreceptors that detect the electrical fields generated by the muscle contractions of its prey; worms, shrimp and even small fish and amphibians.

Underneath this sensitive equipment, the bones of the platypus skull are totally alien to those of their ancestors or any other mammal alive today. The upper jaw, the maxilla, has split apart. Each side curves towards the midline like a pair of pincers. The teeth are gone. Like ghosts they appear briefly in the early stages of a platypus's life, before fading to be reabsorbed and replaced with horny grinding surfaces. With no fossil monotremes to compare, Owen assumed this strange skull was a primitive feature. We now know the platypus evolved from a normal-skulled and fully toothed ancestor. It is exactly because they are so specialised and derived that platypuses are now vulnerable to habitat loss caused by humans.

The platypus's spiny sisters are the echidnas, and they have also evolved startlingly derived skull shapes for a wholly different purpose. There are at least four species of echidna found in Australia, Tasmania

---

* Although interestingly, something similar may have appeared in a group of reptiles called hupehsuchians, in the Triassic. The Triassic is full of weirdos though, as you'll see in Chapter 6.

and New Guinea. Outwardly they somewhat resemble a fat, brunette hedgehog. Like the platypus they have electro- and mechanoreceptors in their snouts, but unlike them, echidnas also employ the more traditional weapons of keen hearing and smell to locate their food. Echidnas are landlubbers – though like most animals they can swim well enough – and are physically specialised for digging up insects, grabbing their helpless invertebrate bodies using an extendable, sticky tongue. As a result, echidna skulls are perfect insect-snuffling tubes. They have no teeth at all, and their bottom jaws (the dentaries) have been reduced to the shape of toothpicks. Despite this, the snout is strong – they can dig and break open hollow logs with it. They've been seen catching insects too big for their tiny mouths – but that's no problem, they simply squash the food with their battering-ram snout and then lick up the nutritious innards.

The other mammals that share the strange tube-like skull-shape of the echidna are fellow insect specialists like the marsupial numbat, or the placental aardvark, anteaters and pangolins. These mammals are vastly separated in terms of biological relatedness, belonging to the three main modern mammal branches. Yet they have independently evolved the same anatomy to find and eat their food. This is called convergent evolution, when two groups of animals arrive at the same adaptive solution in response to their environment. It's a pervasive pattern we see in the fossil record – and one we will return to time and again in our story. This repetition of forms illuminates the lives of extinct creatures. Convergence is exactly what Cuvier had noticed all those years ago in the similarities between unrelated animals living in similar environments, which he wrongly interpreted as evidence against biological change through time.

And so, both genetically and anatomically, the modern monotreme is as advanced as any other mammal alive today. It is true that their ancestors lived in the time of the dinosaurs, but then again, so did ours.

But good stories die hard. In the 1830s Buckland had given the world an explanation for the marsupial jaws from the Secondary rocks of England that fitted perfectly not just with contemporary scientific understanding, but also with pervading cultural and religious views about man's (and white man's at that) superior place in the world. Despite his friend Lyell's attempts to disprove any kind of progression to perfection, the linear narrative of one animal leading to another more complicated one was simple to understand. It made sense to a thoughtful animal that lives a linear life. As a result, Buckland's story of primitive marsupials as the primordial warm-up act for more advanced mammals wriggled deep into the public consciousness and remained there.

Cuvier eventually took a second look at the Stonesfield marsupials. Although he still believed they were marsupials, he now accepted they weren't quite like living opossums, and must belong to an extinct relative. Soon the first mammal genera* of the Mesozoic were christened: *Amphitherium* (the original jaw Buckland had noted), *Phascolotherium*, and a little later, *Amphilestes*. By the middle of the nineteenth century two more Jurassic 'mammals'† had been discovered: another from Stonesfield called *Stereognathus*, and from even older rocks in Germany came *Microlestes* (later renamed *Thomasia*, because the original name had already been claimed for a type of beetle, and once an organism has been scientifically named, the name can't be changed and nothing else can be given the same name). A new English site in Dorset also began producing ancient mammal fossils: the Purbeck Formation. It was younger than Stonesfield, but equally fecund. From the 1830s onwards scientists in France and England began arguing over all of these animals and what their bones meant for ancient mammal life.

By the time Owen published his *Monograph of the Fossil Mammalia of the Mesozoic Formations* in 1871 there were 20 genera, most of them from formations in Britain. He argued that the scarcity of these fossils was probably an accurate reflection of their rarity during the age of reptiles. His summary at the end of his monograph left no one in any doubt of what the great anatomist thought of these animals: 'Mesozoic Mammalian life is, without exception ... low, insignificant in size and power ...'[8] He believed their scarcity and their anatomy supported his

---

* Genera, the plural of genus. All organisms are given a genus and species name, so for humans we are genus *Homo*, species *sapiens*.
† Not all of them were technically mammals, based on our stricter modern terminology of the group, but they were very close sisters. These definitions will become clearer as we go on.

belief in 'the Law of Progress from the General to the Special, from the low to the high ...'[9] Owen definitely considered these first mammals inferior creatures.

Owen's views were based on the tiny amount of fossil material known at the time. These scatterings didn't provide enough information for scientists to realise how different these animals were anatomically from those alive today. But his ideas were also shaped by his stubborn refusal to accept the breakthroughs of his scientific opponents – namely, the theory of evolution. Despite his important scientific contributions, Owen's legacy is sullied by his bitter anti-Darwinism. There was no way he could accept any view held by Darwin or his supporters. When Darwin's friend, comparative anatomist Thomas Henry Huxley, questioned Owen's orthodoxy that Mesozoic mammals were primitive compared to modern marsupials, Owen wrote venomously that 'only a physical defect of vision could fail to discern' that the Mesozoic marsupials were of a 'more generalised type'.[10]

The scarcity of mammals known from the Mesozoic world soon changed thanks to new discoveries in North America. Two men in particular would turn palaeontology from a scientific pursuit to a bitter and very public battlefield. Edward Drinker Cope and Othniel Charles Marsh's public feud over collecting and naming the fossils from the stolen lands of their young nation infamously resulted in deceit, slander, theft and dynamite. Although between them they would produce a significant collection of new fossil animals, it was their outrageous squabbles that particularly brought them into the global spotlight. As a result, the figures they and their cronies painted – of hardy scientific mavericks with moustaches twitching and white skin burned in the sun, braving the wild Badlands with guns and shovels – became a pervasive stereotype in palaeontology. It was a model that generations of scientists would model themselves on. Many still do.

Cope and Marsh are famous for finding the huge bones of dinosaurs such as long-necked sauropod behemoths like *Brontosaurus*.* But both

---

* They were not the first to find them of course. Indigenous Americans had discovered fossil bones long before, integrating their insights into oral traditions that would anticipate later Western scientific understanding. To find out more about this, read *Fossil Legends of the First Americans* by Adrienne Mayor.

men were actually more interested in mammals than dinosaurs. Although they focused on Cenozoic fossils (after the extinction that killed the non-bird dinosaurs), they also amassed an impressive collection of Jurassic mammals. By 1890 the number of mammals from the Mesozoic had more than doubled.

Although mostly represented by teeth and jaws, the material was enough to spark new questions about the diversity and relationships of mammals in the age of reptiles. Unlike other tooth-bearing animal groups such as reptiles, which usually have similarly shaped teeth along the whole jaw, most mammals have specialised dentition: spade-like incisors, pointy canines, bumpy premolars and molars. Another important realisation was that different groups of mammals have similar numbers of each type of tooth, and the shape of their cusps and crests is unique. This meant that teeth, although scattered and rare, could reveal enormous amounts about the relationships between extinct species. It was to become the key to studying fossil mammals.

Mesozoic mammal palaeontology quickly became a science of cusps, tooth counts and dietary speculations. When Henry Fairfield Osborn[*] summarised all known Mesozoic mammals in 1887, he emphasised diet as one of the main ways to classify these enigmatic little beasts: 'these families unite in small groups ... we may divide these sub-groups into carnivorous, omnivorous, insectivorous and herbivorous series'.[11]

But Osborn struggled to fit all the known genera into sensible groupings. He thought some were clearly marsupial, or 'proto-marsupialia', but were 'the Jurassic members ... also to be placed in this order or do they form a distinct order by themselves?'[12] Although mammal teeth are remarkably useful, they have their limitations. As the 1800s drew to a close, palaeontologists were still trying to make sense of Mesozoic mammal families. With only teeth and jaws to go on, the picture remained unclear.

---

[*] One of the big names in American palaeontology, Osborn is less often publicly acknowledged for his racist views about ethnicity and evolution. He shared these views with people like Robert Broom – we'll hear more about Broom and ideas about race in Chapter 7.

It began to dawn on researchers that their assumptions that these earliest mammals were marsupials might not be correct. With Darwin's publication of *On the Origin of Species by Means of Natural Selection* in 1859, zoologists and biologists began examining life on Earth in a completely new light. Previously they had envisaged a progression from reptile to monotremes to marsupial and then placental mammals, in a straight line from simpler to more complex. All of life on Earth could be slotted into this 'great chain of being'. This is where the defunct phrase 'missing link' comes from, referring to links in the chain. But now, researchers began thinking in trees, united by the trunks of common ancestry. The branches were far more complicated, and working out the shape of the tree meant identifying traits that united some groups by shared descent, while distinguishing others by their novel adaptations. Darwin's theory had introduced a mechanism that explained life not as a series of catastrophic replacements, but in terms of the steady modification of groups of animals over time.

By the beginning of the twentieth century, it was understood that Mesozoic mammals from places like Stonesfield were not from modern groups after all, but were the ancient predecessors of the animals we see around us today. Some appeared to belong to the three main modern mammal lineages, while others were clearly something altogether different. It was noted that these creatures also had skeletal traits in common with reptiles – could they have shared an ancestor? Did mammals evolve *from* reptiles? Only a lack of fossils held palaeontologists back as they sought to understand the tree of our furry, milk-producing lineage.

Although the picture was clarifying among scientists, discoveries were slow to be disseminated to the general public. In 1905 the book *Nebula to Man* by Henry Robert Knipe took readers on a sensational poetic journey through geological time. The pages dedicated to the Mesozoic were lavishly colour-illustrated with toothy, bat-like pterosaurs, water-spouting marine reptiles and awkward, gangly dinosaurs. The science was relatively up to date for the time, but the illustrations were the culmination of 80 years of Victorian palaeo-art: the creatures are brutish, slow and crocodilian, as befits their 'primitive' nature. In one image, a blunt-headed pterosaur swoops down on the verdant edge of a lake where a crocodile has just caught a plump Jurassic platypus. A second platypus flees along the grassy bank. In another picture, *Megalosaurus* chows down on an echidna's bum. Many of these illustrations are by the prolific artist Alice Woodward. She and her sisters made huge contributions to science through their art, which was included in books, articles and scientific papers by many of the most important scientists of the day.

Nebula to Man brought the world an epic account of the story of life. When Knipe wrote it, he captured in his text and images a vivid narrative of evolution. Although it included many of the latest discoveries, it was heavily influenced by the idea of progression: primitive Mesozoic platypuses still shared the age of reptiles with the first marsupials. The main theme was life's evolution, understood to take place by Darwin and Wallace's theory of natural selection, but people still held on to the idea of improvement, believing evolution advanced living organisms towards an ultimate goal. The mammal lineage was fated for glory, and the story culminated with recent geological history and the masterful placental mammal race. The pinnacle of this advancement was, naturally, still humankind.

Although this picture was outdated, it was a long time before the wider public was aware that science had moved on. Palaeontologists knew their fossils belonged to groups unknown in the modern world. They had abandoned the pre-diluvian opossums, and there were no more Jurassic platypuses. By the dawn of the twentieth century, scientists had realised that mammals at the time of the dinosaurs were unique, and that they evolved from a lineage with roots even further into the depths of geological time.

But misunderstandings persisted about the elusive origins of mammals and their role in a Mesozoic world seemingly dominated by giant reptiles. It would be another century before popular culture started to catch up with the developing science of mammal evolution. For 150 years Owen's description of Triassic, Jurassic and Cretaceous mammals as nothing more than 'the low and the small; rat-like, shrew-like, forms of the most stupid and unintelligent order of sucklers' continued to resonate with the public, and with most scientists.

We now know this picture of mammal ancestors is a world away from the dynamic creatures that trail-blazed into the modern world.

They were ecological pioneers, and anatomical wizards. They didn't just diverge from their reptile brothers, they ditched them before their evolutionary journey had even begun. They 'ruled the Earth' when dinosaurs weren't even a twinkle in the planet's eye. They were hot-blooded proliferators and dietary innovators, sail-backed show-offs and big-brained micro-ninjas.

It is finally time to hear the *true* story of our mighty mammal origins.

CHAPTER THREE

Like a Hole in the Head

But they are emerging, slowly
Along the shores and the riverbanks
The shingles and the strands
Near Bass Rock and along the Whiteadder
Lungfish and other tetrapods
Crawling out of history

Justin Sales, *Romer's Gap*

Nothing springs fully formed into the world; this isn't a Greek myth. The animals we now call mammals are no exception. Their – our – evolutionary history is tangled together with that of all other life on Earth. But there was a point at which our lineage split away from the rest and began beating its own trail through the jungle of geological time. That point was long before the dinosaurs. It was a time when the continents we know today were joined together. There was only one ocean, Panthalassa, and one landmass, the supercontinent Pangaea. There, on that Janus world, we have our mammal origins – and they lie in our holey skulls.

Around 350 million years ago, Africa and the Americas huddled together in the centre of a plate-tectonic scrum. Antarctica and Australia were attached to them in the south, and together they formed the continent of Gondwana. Meanwhile, North America, Europe and most of Asia – minus India, which was snuggling up to Africa at the time – were attached in the north, comprising the continent of Laurasia. Reaching out from the supercontinent like pinching fingers, parts of China and South-East Asia were strung across the latitudes in an eastern island chain. They enclosed the only other significant body of water on the Earth: the shallow Palaeo-Tethys Ocean, a cradle of coral warmth.

The Earth's equator was straddled by the Central Pangaean mountains, like a belt delineating the hips of Gondwana from the chest of Laurasia. These mountains had formed as the two halves of Pangaea collided, thrusting the Earth's crust upwards in an process geologists have given a suitably erotic name: an orogeny. This mountain-forming event resulted in a chain of peaks that have persisted to the present day: we now know them as the Appalachian Mountains in North America, part of the Atlas in Morocco, and parts of the Scottish Highlands, including the country's highest mountain, Ben Nevis.

The Earth was a cosy 20° Celsius on average (currently it is only about 14° Celsius, but rising fast). Although there were glaciers across the southernmost reaches, most of Pangaea was coated in dense tropical swamp-forests. They were unlike anything we know on Earth today. A journey through these jungles would be an almost unrecognisable experience.

This was the Carboniferous Earth. It is where we began, not only as terrestrial animals, but also as industrial ones.

You are paddling down a tributary of a wide river, curving through the Carboniferous jungles of Scotland. You feel exhilarated by the hot, damp air flooding your lungs. You are near the equator, and you feel strong, heady even – you could run marathons and scale mountains! This is the effect of inhaling the highest concentration of atmospheric oxygen in geological history. Today, the air we breathe is about one-fifth oxygen. Between 300 and 360 million years ago it was one-third oxygen. Blackened scars on the surrounding tree trunks hint at the frequent forest fires that scorch the landscape. Triggered by lightning storms, these fires start and burn readily in such a flammable atmosphere.

Touching one of the charred trunks, you realise the 'trees' are not trees at all. Today we pick up cones beneath the pines, listen to eucalyptus hissing in the wind, stroll under the leaf-littered boughs of oak or push through the wet tangle of a mangrove; all of these forests represent more recent evolutionary groups. The first forests were filled with the giant ancestors and relatives of ferns, mosses and horsetails. We now think of these as mere diminutive understorey inhabitants, lurking lushly in the damp places of the world. But in the first forests, their kind grew to redwood proportions.

You can see the fossilised trunks of these 'trees' in places like Fossil Grove in Glasgow, Scotland. The city is one of the greenest in Europe thanks to a network of public parks. Fossil Grove is situated in an old quarry in Victoria Park, and is one of the metropolis's most under-appreciated treasures. After walking through the lush grounds, replete with mature chestnut trees and garish technicolour flower-beds, visitors follow the path through the rhododendrons and find themselves in a hidden hollow. A small Victorian building sits like a portal at its heart. Inside it you pass through time – and step into Carboniferous Scotland.

Eleven tree trunks emerge from the sandstone floor. They are around a metre in diameter and height, with flat tops that make it appear as though a lumberjack just passed through and felled them. These are 325 million-year-old *Lepidodendron*, an ancient relative of todays' tiny quillworts and club-mosses. The forests of Pangaea in the first half of the Carboniferous were dominated by this plant and its kin, called lycopsids. They are one of the oldest lineages of vascular plants* on the planet. These arboreal ancestors attained heights of over 30 metres (nearly 100ft), and much of the coal that fuelled the industrial revolution came from their fallen boughs.

No carpenter would thank you for putting this material on his workbench: most of their trunks are made of thick outer layers and bark, with almost no wood. The fossil trees in Glasgow are ashen-grey, but their roots reach out from the trunks and dig their fingers into the ground, hinting that they once ran rich with life. Needle-like leaves would have sprouted on every surface. These fell out as the plant grew, leaving behind a crochet-blanket of diamond-shaped scars on their trunks, with only a hairdo of needles on their crown.

The haunting scene of Fossil Grove formed when a muddy flood submerged the bases of the *Lepidodendron*, killing them. They rotted, and left behind empty casts which were later filled with sandy sediment and turned to stone. The fossils were discovered during landscaping of

---

* Vascular plants – as opposed to non-vascular plants - are those that have tissues to transport water from the roots to the leaves, and nutrients from the leaves to the rest of the plant (tissues called xylem and phloem). They also reproduce from sporophytes, which have two sets of chromosomes instead of one, and they have true roots, leaves and stems. Non-vascular plants, like mosses and liverworts, lack these characteristics.

the park in 1887. Recognising their significance, members of the Geological Society of Glasgow suggested they be fully uncovered and left in situ. A building was placed over them for protection, and since 1890 the public and scientists alike have come to transport themselves back to the ancient forests. By the end of the Carboniferous, pines, cypress, gingko and cycads – plants known as gymnosperms – would begin replacing lycopsids as the dominant species, as the climate cooled and dried. But in the early part of this time period, the quillwort-cousin was king.

In the swamp-forest on your journey up the Scottish equatorial tributary, the feeling of wellness you had from all the oxygen soon passes as you discover another repercussion of an O_2-rich world. The fern-riddled undergrowth shudders and sways to the movement of dozens of feet, all belonging to a single creature. Here, there be *monsters*.

We vertebrates were not the first to take advantage of dry land. Fungi and then plants took the first step by approximately 470 million years ago in the Ordovician. Around 50 million years later, arthropods followed them. They thrived for around 70 million years before any backboned animals cottoned on to terrestrial colonisation.

Arthropods are sometimes colloquially referred to as insects, but they actually comprise all segmented creatures with a hard exoskeleton and jointed appendages. Alongside insects this includes crustaceans (such as crabs), arachnids (spiders) and myriapods (centipedes and millipedes). Their many-legged exoskeletons made them ideal terrestrial pioneers: the hard outer casings supported them against gravity out of the water, and locked in vital body moisture. The world was an arthropod's paradise in the Palaeozoic, and they quickly diversified into multiple different forms.

Spiders and millipedes were the first off the starting block. One millipede from 423 million-year-old rocks in Aberdeenshire, *Pneumodesmus*, provides the earliest fossil evidence for air-breathing, through pore-like holes in the exoskeleton called spiracles. These spiracles led inside to the blood and organs, allowing gaseous exchange: oxygen in and carbon dioxide out. It's an efficient system that evolved in different arthropod groups independently through convergent evolution, and persists to this day.

But having a body full of breathing holes poses a problem: they don't just let gases out, they also leak moisture. This is one of the reasons that insects and myriapods are mostly quite small. To grow bigger they need more oxygen, which means larger and more numerous spiracles. But that would cause them to dry out and die. It's no coincidence that one of the largest insects on Earth today, the South American giant longhorn beetle *Titanus giganteus,* which grows as long as an adult human hand, lives in wet tropical rainforests. Research suggests that as insects get bigger, their spiracles and the tubes that connect them internally (called trachea) have to take up increasing space inside the body to do their job effectively. This scaling-up can only go so far, placing a natural limit of about 15 centimetres in body length. That puts *Titanus* at the top end of what's physically possible for land-living arthropods today.

But the natural limit on arthropod body size is different when you have an atmosphere rich in oxygen. In those conditions the gases filtering through the insect body pack a more powerful punch. As you nervously pull the fern fronds aside to find out what's moving through the Carboniferous undergrowth, the source of a hundred footfalls is revealed. Trampling across your path is a millipede the size of a bicycle.

In Laggan harbour, on the Isle of Arran in Scotland, lie traces of these giants. A parallel row of indents in the rock record the scuttling path of *Arthropleura.* This enormous millipede is the largest land invertebrate of all time. Its many footprints look like 'tear here to open' holes on a piece of paper, except that each one is a hand–span apart. Across the Atlantic in Nova Scotia, Canada, similar tracks have been found that are half a *metre* apart. It would certainly be enough to give all but the most enthusiastic myriapodologist the willies.

Falling backwards in surprise, you look up to see what appears to be a sparrowhawk hurtling towards you through a gap in the *Lepidodendron* crowns. But no, it is nothing so inoffensive. As it comes helicoptering past you see it's a dragonfly as long as your arm. As well as being the first on land, arthropods were also the first to exploit the air. Creatures like this early dragonfly relative are the largest flying insects of all time. They were predators, feeding on other insects...

though for a juicy human interloper time-travelling from the future, they might make an exception.

Giant insects up to 10 times the size of those alive today were common in the Carboniferous. Although arthropod fossils have yet to be found in the very earliest rocks from this period, in later deposits they are bountiful and diverse, suggesting they were present and thriving long before. The earliest relatives of mayflies and cockroaches, and spider relatives such as scorpions, along with several insect groups that later became extinct, crawled all over the Carboniferous Earth. It was an entomologist's utopia.

As you sit there cowering in the watery margins of the insect-infested swamp-forests, you finally come across the reason for your visit. In the weed-choked ponds and waterways, something is pushing its way to the surface. It is no bigger than your forearm, and has an elongate body. Two big eyes sit on either side of a long face. Its limbs, which barely lift it off the damp earth to navigate the tangled weeds, jut outwards from the body. It wiggles as it moves towards you, like Dad at a disco. Those limbs end in something that would change the whole direction of life on Earth. They are nature going digital: the first fingers and toes.

In the Carboniferous swamp-forests the first tetrapods followed the invertebrates out of the water. They were the earliest animals on the path to mammaldom – and amphibiandom, and reptiledom, and birddom, indeed these animals are the shared ancestors of all the creatures with backbones and four limbs* on Earth today. Their story is still muddy, but the myriad of fossils now spanning the Early Devonian to Late Carboniferous provide several clear paragraphs – more than enough to piece together the narrative of their emergence.

The clear antecedents of the tetrapods (which literally means 'four-footed') lived entirely in water. They belonged to the group of bony fishes known as lobe-finned fish, or Sarcopterygii. The coelacanths, the lungfish and the tetrapods (which include us of course) together comprise the sarcopterygians – so it is true to say that although we may

---

* Although some tetrapods have since lost two or more limbs, and returned to the water. How terribly ungrateful.

not superficially resemble them, humans are really just highly derived fish (the favourite retort of fish-palaeontologists the world over).*

The group that includes the common ancestor of all tetrapods is called Tetrapodomorpha. Many of their earliest members are known from fossils from Devonian rocks that once sat at the equator, but are now part of Greenland and the Canadian Arctic. These rocks are slightly older than the Carboniferous, at around 360 million years. The fossils have been found by palaeontologists like Jenny Clack and Neil Shubin – folk clearly undeterred by cold, isolation and polar bears. These northern sediments reveal a plethora of possible tetrapod ancestors. We don't know which of them was our *direct* ancestor – this is not something we can ever say for certain about any fossil – but our ancestors were probably similar to animals like *Acanthostega*. It looked like a salamander the length of your arm, with a flat head. Its eyes were placed on the top of its head, looking up to the sky in permanent exasperation. Four limbs stuck out around it, digits wide to form paddles, and tail flattened like a rudder.

It's not just the Arctic that produces tetrapodomorphs, Scotland too had its own. *Elginerpeton*, found near the town of Elgin in Moray (a place we'll return to later), was similar to *Acanthostega* and other early tetrapod ancestors and cousins. They share a recognisable body plan that formed the starting point for all vertebrates that walk the Earth today.

Surprisingly, *Acanthostega* had *eight* fingers and toes on each limb, while other tetrapodomorphs had seven. It wasn't until later that five became the magic digital number. This was probably for practical reasons during weight-bearing, because too many fingers may have been cumbersome, reducing the flexibility of the wrist or ankle and making it harder to move on land. *Acanthostega* didn't have to worry about its weight, because it was probably fully aquatic. Like the modern Australian lungfish, it had lungs as well as gills. This was likely an adaptation to survive in shallow, oxygen-poor water.

Limbs and fingers didn't develop so that lobe-finned fish could pull themselves out of the water; evolution doesn't work with an end-goal

---

* This truth is superbly explored in Neil Shubin's *Your Inner Fish*, and I urge everyone to read it to find out exactly how our lobe-finned ancestry expresses itself in our bodies to this day.

in mind. Instead, limbs allowed these animals to navigate *within* the water, using them as paddles to push aside dense underwater foliage. By analysing the angle of the legs to the body, and the structure of their shoulders and hips, palaeontologists now realise that early tetrapodomorphs would have struggled to support their own weight without the buoyancy provided by water. But with air-breathing lungs and a four-limbed body plan both in place, it wasn't long before these fingersome critters used their adaptations to access drier abodes.

At the close of the Devonian and the beginning of the Carboniferous, there is a gap in the fossil record. It has a name: Romer's Gap, after Alfred Sherwood Romer, a palaeontologist from the United States fascinated with the evolution of vertebrates. Romer was particularly obsessed with the 'fish to tetrapod transition': the transformative evolutionary journey of one group of lobe-finned fish to become the progenitors of all four-limbed animals. His stunningly detailed books – published between 1930 and 1970 – cover the structure of all living vertebrates and are so meticulous and well illustrated that they remain vital texts to this day.

Romer noticed that there was a point in the tetrapodomorph fossil record where we simply didn't have any fossil remains to tell their story – a gap in our knowledge. This absence of fossils was named after him by later researchers. From 375 to 360 million years ago, at the end of the Devonian, two mass extinctions decimated life on Earth. In the following 15 million years (the start of the Carboniferous) the fossil record goes strangely quiet. It has been suggested that an unusually low concentration of oxygen in Earth's atmosphere at that time may have reduced rates of fossilisation, but there may also have simply been fewer animals around to become preserved. This is Romer's Gap. After it, tetrapods are diverse landlubbers able to support their own weight out of water without batting an eyelid.[*]

---

[*] It's not entirely certain whether they would have had eyelids. I asked Stig Walsh (see Chapter 1), who studies them, and he said 'my feeling is that they wouldn't have needed any sort of moisturising structure to protect the eye. At some point something like that would have had to have evolved ... to be honest I've never thought about it ...' I sense a novel research grant proposal on the horizon.

For a long time we didn't know how this water-to-land transition had happened, but recently the gap has started to be filled in. Many of the newest finds come from the borders of Scotland, where scientists working as part of the TW:eed project (Tetrapod World: early evolution and diversification) have been digging up new early terrestrial vertebrates. The work was driven by pioneering palaeontologist Jenny Clack and colleagues, and among their discoveries was *Aytonerpeton microps,* 'the creeping one from Ayton with the small face' (Ayton being the Scottish parish in which she was found).

With nippy teeth and a chin full of dimples, *Aytonerpeton* – or 'Tiny' as she* was christened by the team – has a skull just five centimetres long. In other words she was small, certainly when compared with some of the other tetrapods from the time period. Her rows of sharp little teeth probably snatched up the copious invertebrates that surrounded her. The success of arthropods would have provided a tempting menu for early tetrapods venturing from the waterways.

It is, of course, easy to slip into language that suggests evolution had a goal in mind when the first backboned animals took to the land. Of course, there is no end-goal to evolution's journey. It is random, the route forged by happenstance. Divergences occur when unexpected mutations or behaviours turn out to be useful at any moment in time. This uncaring randomness frightens some people – as evidenced by the anti-evolutionary arguments given by some religious devotees. But it is the serendipity of it all that makes evolution so utterly glorious.

Our tetrapod ancestors didn't evolve any of their adaptations *for* life on land, but the adaptations they already had turned out to be useful platforms from which to launch themselves out there. This phenomenon is called exaptation by scientists, and it is an important feature of how evolution works. A trait that evolves for one purpose is re-tooled to serve another. It wasn't remotely inevitable that Tiny would stride on her way through the Carboniferous Scottish undergrowth, let alone that a strange group of animals that puts all its

---

* I like to think she's a *she*, though there's absolutely no scientific reason to think so.

squishy bits on the outside (crazy idea, just ask an arthropod) would ever amount to anything.

And so this foray through the swamp-forests has brought us to our origins – arguably one of many origins or forks in the trunk of the vertebrate tree of life. The creature you met earlier in the wet undergrowth is *Westlothiana*, named for the region in which the fossil was discovered, West Lothian in the Scottish Borders. We could trace our evolutionary history back further, slithering into the brackish estuaries in search of the common ancestor between lobe-finned sarcopterygian fish and their ray-finned sisters (the actinopterygians, which include the rest of the bony fish on the planet). Or indeed further still, to the insanity of the Ediacaran where the first multi-cellular organisms included flattened blobs that looked like hernia support cushions. But *Westlothiana* is as good a place as any to begin our mammal tale. It was not only one of the first animals on land, but features in the skeleton also tell us that it is related to the group we belong to, the amniotes.

As the Carboniferous period continued, disaster struck those sultry lycopsid swamp-forests. An event dubbed the Carboniferous Rainforest Collapse felled the wetland trees. It's unclear what caused the collapse, but evidence suggests it was a change in climate, driven by the eruption of volcanoes in what is now north-west Europe. There were still forests, but they were different after that: fragmented, and filled with gymnosperms. Their cones and seeds are familiar to us now as conifer trees, the unusual and endangered gingko, and the palm-like cycads much beloved of Victorian botanical collections. Beneath their branches those ambiguous tetrapodomorphs finally came into focus as full-blown, four-footed land animals. There was no doubt any more, these were the progenitors of the two great lineages of vertebrate life on land: amniotes and anamniotes.

As the climate altered, one group of animals retained many of the prototype evolutionary outlines of the first tetrapodomorphs. They continued to rely on water for reproduction, and as a safe place to keep their eggs moist and provide them with oxygen. This group is called the anamniotes, and their need for moisture meant the drier world of the Late Carboniferous was a challenging place. But they hung on: this group survives today in the form of frogs, salamanders and caecilians.*

For the rest of the tetrapods however, two novel adaptations meant they were able to break their reliance on water, severing ties to the depths completely. The first adaptation, which was long given precedence, also gave them their name: the amniotes. It comes from the fluid-filled membrane called an amnion that encloses their developing embryos. The amnion evolved from the jelly-like outer layers of egg you can see around the frogspawn in your pond. In amphibians, this jelly lets waste and gases pass between the spawn and water, but in amniotes the amnion performs the same functions *out* of water. With two other membranes, the chorion and allantois, the amnion forms a cushion of membranes enclosing the developing embryo in its own portable pond: the amniotic fluid. All of this is placed inside an egg-shell (and much later, in some animals, occurs inside the mother's body). It is a feature amphibian and fish eggs lack. Thanks to this eggy innovation, the amniotes could raise their young away from their riparian homelands.

But there is another change in the amniote body that was arguably even more crucial to their transition. As vertebrates left the water, they took advantage of the fact that air has up to 30 times more available oxygen than water. Fish have to flush huge volumes of water over their gills to breathe, but on dry land it is thought that the first tetrapods employed something called buccal pumping. It's a kind of bellows mechanism, created by lifting and dropping the oral cavity. This also explains why a lot of these first tetrapods have wide flat heads that look a little as if they've been stepped on. The shape creates a bigger oral cavity. Basically the first tetrapods were a bunch of mouth-breathers.

Amniotes, however, stuck their necks out, quite literally. Examining the differences in rib mobility in these animals, palaeontologists Christine Janis and Julia Keller noticed in 2001 that the ribs became more mobile in the first amniotes.[1] Many of them had narrower heads

---

* Although we often call them amphibians, it is more correct to call them, and their extinct relatives that share a common ancestor, the lissamphibians. Amphibians are the more recent, living branch of this family.

and longer necks. They realised these changes were all connected, and reflected a change in breathing mode, from mouth–based buccal pumping to costal breathing, using chest muscles. This would have a profound impact on the history of life on Earth. Not only did it mean that they were more efficient breathers, but amniotes were no longer using their stout ribs to maintain their posture. They could stand more upright and elongate their necks – something that would have hindered a buccal pumper from getting enough air into the lungs. With the mouth no longer employed in ventilation, amniotes were free to re-use some of their skull and jaw muscles for new types of feeding. For the first time they could eat vegetation, a diet that required skilled biting at the front of the jaw. With a chesty sigh, amniotes strode away from the water's edge, laid their eggs and took on the world.

At most points in our evolutionary history all we have to go on to understand evolution are the bones. It is clear from the fossil record that amniotes are one of a complex array of animals living alongside one another at this time in Earth's history, all experimenting with life on land. Gradually, this lineage accumulated changes in the skeleton that helped them survive out of water: stronger vertebrae, larger limbs, the restructuring of the ankle to support their body weight and optimise foot movements for walking. Palaeontologists closely examine and chart such changes, observing which ones provide the tell-tale mark of a distinct group. Such features are called synapomorphies, and they distinguish one group from another.

The synapomorphies of amniotes require a keen eye. In the skull, they include the arrangement of the bones: a bone called the frontal widened and formed part of the orbit of the eye. Inside the mouth, part of the palate, the roof of the mouth, had a flange covered in teeth reaching towards the back of the throat. In the shoulder, the bones become more complicated (including the development of two coracoids, which in mammals are now part of our shoulder blades), likely linked to the change in limb-use with life on land. Where the shoulder bones grew more complex, the ankle and wrist simplified, and multiple bones fused to form the astragalus (part of the ankle).

Other features paint an intriguing picture of these animals. For example they had no specialised ear for hearing out of the water, so theirs was a world of heady vibrations. In some early amniotes the structure of the skull appears flexible, and lacks the kind of muscle attachments we associate with precise biting, suggesting they weren't able to eat anything that couldn't be swallowed more or less whole. Some may have continued to feed in water. There are also aspects of their biology that remain hidden from us, such as the keratinisation of their skin, or how the amniotic egg evolved. These are soft-tissue features, and unlike bones they seldom leave any trace in the rock record. We can infer their presence through common ancestry, but for most soft-tissue features there are no fossils to prove their presence or absence conclusively.

We now come to the final fork in the tetrapod tale – after this, we hit the mammal highway. Around 300 million years ago, when the lycopsids were falling and the seeds of the first gymnosperms sprouted their claim on the forest floor, our own lineage had already parted ways with our cousins, the reptiles. It is a common misconception that mammals evolved *from* reptiles. We now know this is not remotely true. But mammals and reptiles do share a common ancestor. This first amniote tetrapod was neither mammal nor reptile – neither of those groups had evolved yet. In the Carboniferous, our last common ancestor with turtles, crocodiles, dinosaurs, birds and lizards said goodbye, and set off into the evolutionary sunset.

The amniote tetrapods were cleaved into two mighty lineages: Synapsida and Sauropsida.* These Romulus and Remus groups looked alike back then, you would struggle to tell them apart at a glance. Traditionally though, we have recognised them in the fossil record by one feature in particular: the number of holes in their heads.

The synapsids include us and all our mammal brothers and sisters, as well as an incredible host of extinct creatures that we will meet in chapters to come. The sauropsids (reptiles), on the other hand, are arguably the more successful of the great tetrapod houses – if you discount the so-called 'success' of humankind. From equally humble

---

* There are arguments about using the term Reptilia or Sauropsida when referring to this branch. The relationships within this tetrapod branch remain hard to resolve. The debate is outside the scope of this book, so I'll use the terms interchangeably here.

beginnings Sauropsida has given rise to a startling diversity of forms, from turtles to pterosaurs, lizards and tuatara,* ichthyosaurs to crocodiles, and of course dinosaurs – who have received more than their fair share of attention. Birds, the living descendants of dinosaurs, are twice as speciose as mammals, setting us all a-twitch with their diversity. But there are plenty of books outlining the reptile evolutionary journey, especially in the Mesozoic.† I needn't tire you with that tale here. Let's focus on their sister-group, for they are the protagonists of this particular evolutionary tale. They are the synapsids.

There are currently several strong candidates for the earliest synapsid, the lineage of four-limbed animals that includes mammals and their relatives. All of them have been found in the Carboniferous rocks of Nova Scotia. It would seem our most ancient synapsid ancestors were Canadians.‡

The latest fossil believed to be from our lineage is *Asaphestera*. The next sounds more like a preventative medication than an animal: *Protoclepsydrops*. Then there is *Archaeothyris* (perhaps some kind of throat malady?) and *Echinerpeton* (Scottish slang, for something naughty no doubt). The names hint at the difficulty their scattered and fragmentary fossils pose for the scientists who study them.

Asaphestera has only recently joined our mammal-line ranks. The name means 'less distinct one', which is appropriate because it was so indistinct, it was originally lumped in with the bones of several unrelated animals. In May 2020 a study[2] came out by a group of scientists from Canada and Germany who had taken a second look at the specimen. They realised the bones of its small wide skull were arranged in a pattern associated with the earliest synapsids. If this is

---

* The tuatara (*Sphenodon*) is the only surviving member of a group of reptiles called rhynchocephalians. It is endemic to New Zealand.

† I recommend *The Rise of Reptiles* by Dr Sues – not the gloriously absurd children's writer, but the German-American palaeontologist. Admittedly, he revels in equally glorious absurdity. And cats, he LOVES cats.

‡ I jest of course. Canada has the oldest known fossil synapsids, but that doesn't mean they weren't present elsewhere in the world at the time, we just haven't found their fossils yet.

the case, *Asaphestera* is one of the oldest stem-mammals (as in the stem of the family tree).

From the same rocks comes *Protoclepsydrops*, which means 'first *Clepsydrops*'. The slightly younger fossil of *Clepsydrops* is yet another Carboniferous Canadian whose vertebrae looked like an hourglass, or Greek klepsydra.* Although it is clear from its skeleton that *Clepsydrops* is a synapsid, opinion remains divided over *Protoclepsydrops*. Its fossils comprise only a few vertebrae and upper arm bones, but they seem to resemble other early synapsids in shape. So this cough-medicine animal may indeed be among the very first of our line.

Archaeothyris means 'ancient window' in Greek, and it's a poetic choice. This fossil provides not only a glimpse of the ancient amniote past, but outlines a defining feature of our synapsid family. We, along with our shared relatives all the way back to the Carboniferous, have a single hole in each side of our skulls called a temporal fenestra. In anatomy, a fenestra (meaning 'window' in Latin) refers to any hole in a bone, but this particular hole is pivotally different. It lies just behind the eye, and synapsids have just one of them on each side. Synapsida means 'one arch', because this single fenestra creates a single arch in the skull bones.

Our temporal fenestra is rimmed by the temporal, squamosal and postorbital bones of the skull.† You can feel your own fenestra by placing your fingers in the hollow behind your eye, above your cheekbone. Now clench and unclench your jaw, and you should feel the muscles running through your temporal fenestra. This opening provides attachment areas for the muscles that open and close the mouth, so it could be that different arrangements of holes in early tetrapods are linked to different ways of biting and feeding.

Recent research has shown that the pattern of how many holes there are in the skulls of reptile lineages is more complicated than

---

* You have to marvel at what classically educated Western palaeontologists used to think was a good visual reference for naming a species. But then, what will people of the future make of recent scientific names derived from film franchises and death metal bands?

† In early synapsids, it is the jugal, squamosal and postorbital bones that create this fenestra. The bones move around a bit through synapsid evolution, but the effect is the same.

Figure 1 Diagrammatic skull of a synapsid (left, mammal-line) and diapsid (right, reptile-line), showing temporal fenestrae.

traditionally thought, with fenestrae being acquired and lost again in multiple groups. Most reptiles are diapsids with two holes, but turtles are anapsids and have none. Early groups may have gained and lost one or more holes through their history. In Synapsida however, the single fenestra is almost unequivocal.* The main split between synapsids and sauropsids is clear-cut and our single skull-hole on each side is a consistent defining feature. *Archaeothyris*'s 'window' is a window into the skull and into our own past, a clear indicator that it belongs at the base of our tree.

Echinerpeton, on the other hand, rather unhelpfully means 'spiny lizard'. Scientific names, once given, cannot be changed even if their meaning is subsequently proven incorrect or misleading. This poses a problem for modern palaeontologists doing their best to make clear the crowbar separation between synapsids and their reptile cousins. Although *Echinerpeton* belongs to the mammal line, it is unfortunately named a lizard in perpetuity.

As well as scientific names, colloquial ones can be infuriatingly diehard. For synapsids, the once commonly used term for these ancestors of mammals is 'mammal-like reptiles'. I cringe as I write it. Some of the best books on the origin of mammals use this, keeping the term alive in new generations of students and the public. It is a relic from an extinct terminology, but it's easy to find

---

*Rather unhelpfully, one group of synapsids had a very small second hole in the skull. It was only very small, but it just goes to show that there is an annoying exception for every rule, which is why biological sciences are always so much more complicated than they at first appear.

LIKE A HOLE IN THE HEAD

yourself falling back on it as a familiar touchstone, which only keeps the misnomer going. In truth, we need the term 'mammal-like reptiles' about as much as a hole in the head.*

You might think I'm over-reacting, but I'm not. 'Mammal-like reptiles' reflects a fundamental wrongness about where we and all our milky brothers and sisters come from. It's the taxonomic equivalent of insulting your mother. The truth of mammal origins is so much more fantastic, and knowing it transforms how we see ourselves and the animals we share our planet with.

In the quest to understand how evolution works, mammals that we are, we have tended to search for mammalness in the fossil record, to trace our lineage back to its genesis. We are creatures that perceive time as a linear experience with a start, middle and end, and we structure our stories about the world accordingly. Our language is excellently utilised for delighting us with these tales, but it is also riddled with ambiguities that are less than ideal for scientific accuracy. This is why scientific language seems opaque to those who don't habitually use it. A very specific terminology is needed to ensure every researcher is discussing exactly the same thing, and prevent misunderstanding. This can look like pedantry, but it is not. Attention to wording has a purpose, a clarity.

When it comes to evolution, it is really hard to use the correct words rather than the best narrative ones. We talk about animals that 'turn into' or 'become' other animals through the process of evolution – I'll probably do it multiple times in this book. These are *Just So* stories.†
This happens because we view evolution from back to front, comparing everything to what lives today. This approach has its uses – palaeontology is founded on comparative anatomy, and in later chapters we'll see how such comparisons in the fields of biomechanics and ecology can tell us how extinct animals might have lived. But it also has the consequence that we see inevitability in what is actually just randomness. It means

---

* One hole on either side, that is ....
† Referring to Rudyard Kipling's children's classic from 1902, which explained the origin of different animals' characteristics in a Lamarckian narrative. In other words, each animal chooses to develop a feature for its own benefit. They may not be evolutionarily correct, but they make good bedtime stories.

we miss important distinctions like the one we must make now, before we go any further, between the outward appearance of reptiles and their shared common ancestors with mammals.

Reconstructing *Asaphestera, Protoclepsydrops, Archaeothyris* or *Echinerpeton*, they would have looked rather like lizards. They were all small, not bigger than your forearm. Their legs stuck out to the side and their long pointy tails would have swayed and dragged as they waddled along in the characteristic 'reptilian' wiggle (reptiles still move more like fish, with side-to-side motions of the body). There would have been no fur or feathers on these first synapsids, just a tough textured surface to the skin, locking all their precious moisture in. Their long lipless mouths were filled with rows of simple pointed teeth to crunch down on insects or fish.

A suite of subtle skeletal features define the earliest synapsids. Aside from the main synapomorphy of having one large hole in the skull on each side of the head, the bone that forms part of the back of the eye region became broader and tilted, and the septomaxilla bone in the nose enlarged. Outwardly however, they looked like reptiles. They probably acted like reptiles, and so did the first sauropsids that lived alongside them. But technically *none* of them are really reptiles.

Despite appearances, the early amniote tetrapods – the synapsids and sauropsids – were not 'reptile-like'. Instead, we ought to say that reptiles today are 'early amniote-tetrapod-like'. Most modern reptile groups (particularly lizards) have retained a lot of characteristics that are reminiscent of their forebears; they've not *superficially* changed that much, they're just a bit retro.* Meanwhile the mammals have changed radically and obviously (as have birds). This change is clear even to the casual observer. No one ever mistook a cow for a chameleon, or an eagle for an iguana. But one might easily mistake a cow's early ancestor in the Carboniferous for an eagle's early ancestor, because they all looked pretty similar. This is where a detailed understanding of anatomy is crucial. Anatomy is the only way to tell the difference between them and piece together how they emerged into distinct groups.

---

* Of course in truth reptiles have changed a great deal, anatomically and genetically. But that's a story for a different author …

The repercussion of this similarity was that the first palaeontologists and anatomists traced mammals back in time, and perceived their fossils as looking increasingly 'reptilian'. They therefore assumed mammals evolved from a branch of reptiles, and so the term 'mammal-like reptiles' was born. With more fossils and increasingly detailed analyses, we now know mammals did not evolve from reptiles at all. The reptile-ness of the early synapsids was just a hangover from the early amniote tetrapod body plan, which they share with reptiles.

↓↓↓

The split between mammals and reptiles, synapsids and sauropsids, runs deep. They were cleaved in the Carboniferous, in a world of pioneer terrestrialism. Most of their fossils have been found in rocks that once lay near the equator, in the hot humid forests of the ancient world, suggesting they were not yet able to live in drier or colder environments. Like a parent, only a palaeontologist could tell these founding twins apart, but their differences are fundamental to who they are.

It is the fossilised forests of the Carboniferous that fostered our early evolution – and more recently, drove our industrial revolution. Their persistence over millions of years, then their collapse and fossilisation, produced the deep veins of coal that run around the world. The exploitation of these resources is intimately tied to the foundation of Western empires and colonialism. Burning fossil fuel has triggered the rapid anthropogenic climate change that is now altering millions of lives, particularly in the developing world, and will continue to do so in the coming centuries.

In a way, not only does the Carboniferous mark our worldly beginnings, but it may also be the source of our ultimate destruction.

CHAPTER FOUR

The First Age of Mammals

All these groups are very distinct now, and we naturally turn back to ancient times to ask how they first started each upon their own road. But when we do this, we meet a history so strange that it makes us long to open the great book of Nature still further, and by ransacking the crust of the earth in all countries to try and find the explanation…

Arabella Buckley, *Winners in Life's Race*

My least favourite bits of history class were about royalty and battles. Our classes were almost all run by middle-aged white men who taught us about other middle-aged white men and the wars they waged. These wars were usually at the whim of richer white men who, along with their rich white wives, grew empires, invaded one another, and collapsed again … only to be replaced by new empires, kings and the men they sent to war. Military history and kingdoms really float boats, but the emphasis on the glory of winning is a rather shallow take on history. It is also a very misogynist one, usually told from a Eurocentric perspective. What were regular people doing? What was happening in Australia? Or Argentina? Surely the past isn't just a list of kings and maps of their fluctuating borders?

Until now men have tended to write the history, and be written about.[*, 1, 2] The same has been true of evolutionary history. The language of conquest permeates how we talk about evolution: animals *dominate*, they are *kings* of the jungle, they *rule* the seas/land/skies, and all other groups sit in their shadows like vassals. How tedious it is always talking about the rise and fall of life as though it was a series of oppressive empires.

It is true, however, that groups of animals have had peaks and troughs of being the most diverse and populous on the Earth. We call the Devonian 'the age of fishes', because there were a heck of a lot of them, both in the sea and in the fossil record, and they were doing

---

[*] I'm generalising, but it's mostly true.

such darned interesting things while most other animals weren't even off the starting block. For example there were 'armoured' fishes, another interesting choice of descriptor. As with human history, our portrayal of the history of evolution is biased, and reflects the culture of those who tell it (and I'm no exception). When it comes to our perception of 'dominant' life, it is skewed by what we focus on (or gloss over), and in some cases it's just incorrect.

For example, we are now living in the 'age of mammals'. You may have guessed it already, but I'm pretty fond of mammals. However, to say that they are the dominant life of the last 66 million years requires a bit of tunnel vision. There are over 5,500 species of mammals alive today, versus a staggering 18,000 species of bird, and over 35,000 fishes.* That's just vertebrates. Among insects, beetles alone tally over one and a half million species. The reasoning behind our time's moniker is that mammals include the largest vertebrates on Earth, and we are disproportionately focused on size – in light of the male bias in history and science, make of that what you will. They are also exceptionally diverse in terms of body shape and how they live in their myriad environments. And to be honest, we just have a big old soft spot for them, our fluffy brothers and sisters.

But the true 'age of mammals' actually happened a very long time ago. You might have missed it on a casual perusal of evolutionary textbooks. There are almost no titles on the subject in the popular science section of your book store (until now). We tend to skip it, much as my history teachers went straight from the Stone Age to the First World War, ignoring the great inventions of the Chinese civilisation or the mind-blowing exploration of the Pacific islands by the Melanesians along the way. We go straight from fish to dinosaurs, skipping casually across a 250 million-year gap in the story. At most, there's a passing hat-tip to an erroneous 'mammal-like reptile' – sometimes accidentally plopped in with the dinosaurs – before we get stuck into the meat of the age of reptiles.

Where are those missing millions of years? What were animals up to, between crawling onto land and turning into real-life Godzillas?

---

* On 15 June 2020 it stood at 35,519 species of fish, according to *Eschmeyer's Catalog of Fishes*.

Let me cut back the foliage and reveal the first great proliferation of vertebrate animals. These are the creatures that streamlined vegetarianism and gave apex predation its test drive. The shocking diversity that synapsids accomplished in size and shape, and the adaptations they evolved to live their best life, defines this hidden period of geological time as the true first age of 'mammals'.

↓↓↓

We bumped ungracefully along the dirt track road, pothole craters causing the car to jolt and scrape disconcertingly towards the sea. We parked at the end of the track, followed by another two cars and a 4X4. They were driven by staff from NatureScot and Elgin Museum.* Stepping out, we were met by a typical Scottish winter day: the sun made pretences to shine, but the air was sharp as glass, sending shards from the North Sea sliding up our sleeves to cut to the bone.

I had joined the group alongside Nicholas Fraser, Keeper of Natural Sciences at National Museums Scotland in Edinburgh and one of my PhD supervisors. Fraser specialises in the palaeontology of the Triassic, a time when life restarted like a crashed computer after the most devastating mass extinction of all time (more on that later). The quarries around Elgin in Moray, north-east Scotland, had historically yielded some of the first specimens of Permian and Triassic animals in the world, earning them protected status. As a Triassic worker and Keeper of Scotland's largest museum, Fraser had been asked to come to Elgin to help assess the need for continued palaeontological protection of the sandstone quarries of Moray.

We set off along a track towards the sea. On either side, fearsome walls of gorse hid the view and deterred straying feet. Eventually we came to the coast and below us the cliffs dropped onto sandstone shelves that slid into the churning sea. The wind was powerful, snatching my breath. My nose tingled with brine and eggy seaweeds. We could just about discern the opposite coast of the Moray Firth over 30 miles away, shadowed under threatening clouds. The water between lay dark and restless, punctured by the occasional hopeful seafarer.

---

* One of the oldest independent museums in the UK, having opened in 1843. It is well worth a visit for the comprehensive fossil collection, along with local historical items.

Just as I was wondering what we'd come so far and shivered so much to see, we rounded a corner and the ground opened up. Below us was a quarry pit over 100 metres across. The back wall was a ragged cliff, pin-striped with boreholes where the quarry workers had drilled down from above and blasted the face away. To the left squatted three rusted huts, their edges so gnawed by salt and time that it seemed a gentle touch might collapse them into flat-packs. In the heart of the quarry stood boulders, metres across in every dimension. Beyond them lay heaps of rubble with sandy pathways cleared between, and wide shimmering puddles.

This was Clashach, a working sandstone quarry and source of building material for sites across Scotland. It is also one of Scotland's greatest fossil footprint localities, and one of the best Permian-aged footprint sites in the world.

The earth has been flayed open at Clashach, revealing pale yellow guts. The 260 million-year-old sandstone was quite soft – one of the reasons the quarry lay unused for years. But in the 1990s when the Museum of Scotland and Royal Museum merged their collections in the National Museum of Scotland building in Edinburgh, Clashach sandstone was chosen for the cladding for their new building. Controversy over the unusual architectural design was fodder for newspapers across Scotland, but what the journalists barely mentioned was that renewed work at Clashach had produced more than just divisive building materials. Blasting and extracting in this coastal quarry exposed new footprints belonging to long-extinct animals, peppering the petrified dunes.

Now only small-scale quarrying takes place, mainly to produce gravel.* The company leasing the rights wanted to expand into new sections, and we were there to assist in the decision-making process for their expansion. We had to figure out if Clashach Quarry still needed formal protection, or if the fossil trackways had long since been exhausted.

---

* Some Clashach stone has been used in the Basilica of the Sagrada Família, a dazzling temple in Barcelona designed by the flamboyant Antoni Gaudí. The structure is 135 years in the making, and counting. Clashach stone was chosen for its similarity to local Spanish Montjuïc stone.

The quarry was empty as our group of seven strode in. Our feet sank into a thick layer of waterlogged sand – runoff from the slowly disintegrating stone. We reached the large boulders in the centre of the quarry's amphitheatre. The staff from Elgin Museum explained that boulders were often put aside by the workmen because they contained footprints. We clambered around and over them, searching. My fingerprints rasped the sand grains. Slight bumps, not more than a midge-bite.

Further in lay more slabs of rock the size of double beds. As we approached them the sun sklent over the shimmering peachy surfaces. Permian paw-prints were suddenly brought into sharp relief. It was like a late evening on the beach at low tide, when dogs and browsing shore-birds have left their trails in the damp sand. At first the Clashach prints seemed like no more than alternating indents, but then we began to pick out claws: digging in, dragging aside. Some prints were the size of your fist, others as small as thumb-prints. There were toes. Even the more cryptic marks began to make sense as I followed the rhythm of a saunter, or the shuffle as two tracks met, then diverged.

The ghosts of Permian beasts were walking all around us.

The Permian was not named for the hairdo of course, but for the region of Russia. Outside a school in the city of Perm (administrative capital of Perm), a rather uninspiring lump of grey rock bears a plaque with the inscription:

> To Roderick Impey Murchison, Scottish geologist, explorer of Perm Krai,
>
> who gives to the last period of the Palaeozoic era the name of Permian.[3]

A memorial to Murchison's geological endeavours also adorns the bank of the Chusovaya River, a tributary of the surging Volga that runs through the Perm region, with its roots in the Ural Mountains. This memorial includes an image of the teeth of the Permian-aged shark *Helicoprion*, which grew in a bizarre spiral in its lower jaw, as though *Jaws* had a baby with a tin-opener. Murchison is one of

Scotland's most well-known geologists (we've produced quite a few) and he left his fingerprints all over those distant mountains and their people, thanks to his time spent geologising across Russia in the 1840s with his pals. We have him to thank for adorning this time period with its moniker.

The Permian Earth, 299–252 million years ago, retained a somewhat Carboniferous face. The continents were still scrunched together, but the sea-level drop that had begun in the preceding era continued, exposing huge areas of land that had previously been inland seas. This world was distributed to extremes – half water, half land – and produced extreme environments. Towards the poles lay cool temperate forests, the leftovers from the dense woodland world of the earlier Carboniferous. But these had dried out and were filled with conifers and seed ferns. The poles themselves were eventually ice-free thanks to unhindered ocean circulation and the steady warming and drying out of the land. Along the coast of the Palaeo-Tethys Ocean, monsoons seasonally deluged onto the landscape, producing lush and humid environments.

Cut off from the sea, the huge expanses of inland supercontinent at the equator and mid-latitudes reached sizzling temperatures of over 40° Celsius. These were the parched interiors of Gondwana and Laurasia. Scotland was cradled there, in the heart of northern Pangaea. Part of an arid, wind-swept desert like the Namib or Sahara today, it would have been a harsh region to survive in. Yet fossils from these ancient dunes suggest life not only adapted, it thrived.

The complex ecosystems of the Permian are often overlooked when people tell the great story of evolution. Ask a non-palaeontologist to tell you about it, and they'll draw a 50 million-year blank. But the Permian wasn't just another time period, it was another *Earth*. The animals that populated it were evolution's first play-through. These Atlantean beasts included some of the first tetrapods to exploit herbivory. They grew to a gigantic size, positively Pantagruelian, and sprouted horned defences long before *Triceratops* jumped on the bandwagon. The first fast-running hunters stalked the landscape with sabre-teeth. Those prey that couldn't grow, shrank, and dug deep into the earth for safety.

You may think that a planet full of dinosaurs is hard to imagine, but the strange figures of the Permian dreamtime represent a true alien world. Yet it was one that foreshadowed our own. This dinosaur-less landscape was the playground of our oldest and mightiest mammal ancestors.

Nothing could feel further from the hot Permian desert than a Scottish quarry in November. I scrabbled over the sandstone debris in my woolly layers, looking for fossils. The study of footprints is part of the wider study of trace fossils, called ichnology. This includes footprints, burrows and droppings – not the fossil animals themselves, but their marks upon the world. The ichnology of footprints began in Scotland with the discoveries of tracks made by animals with backbones and four limbs, found in quarries. They belonged to our synapsid ancestors and their contemporaries, who lived in the sizzling Pangaean heartland.

Dumfries, in the hills of south-west Scotland, bears rocks older than those in Elgin, but from the same time period. In the early nineteenth century quarry workers at a site called Corncockle Muir began to notice a number of marks in the reddish sandstone beds they were cutting for building material. Several of these tracks were collected by curious locals, some even being incorporated in the walls of a summer-house as a curiosity.[4]

By 1828 the strange markings caught the attention of Victorian naturalists, including William Buckland. It was just four years since he had described and named the first dinosaur, *Megalosaurus*, and mentioned its less conspicuous but no less important companion, the Mesozoic mammal *Amphitherium*. Buckland had been made aware of the footprints in Southern Scotland after being sent casts by the Reverend Henry Duncan of Ruthwell parish, who was a minister and geologist. 'The professor [was] ... fully convinced by them that the rock, while in a soft state, had been traversed by living quadrupeds ...'[5]

Buckland asked Reverend Duncan to send him more specimens, 'at any expense'. So Duncan and his friend James Grierson visited the quarry at Corncockle. Grierson described what he saw:

The great number of the impressions in uninterrupted continuity – their equidistance from each other – the outward direction of the toes – the grazing of the foot along the surface before it was firmly planted – the deeper impression made by the toe than the heel ... the sharp, and well-defined marks of the three claws of the animal's foot – are circumstances which immediately arrest the attention of the observer, and force him to acknowledge that they admit of only one explanation.

The explanation was that four-legged life had walked on those sandy dunes 'before the flood'.

In the same year as Grierson's account, Duncan gave an address on the prints.[6] Where Grierson had provided a brief overview, Duncan eloquently described these extraordinary fossilised footprints, 'varying from the size of a hare's paw to that of a foal's hoof'. He went into much greater detail, observing not only the size and the distances between prints, but their depth and shape. In the earliest attempt of its kind, Duncan used the shape, or morphology, of the prints to suggest the behaviour and proportions of the track-makers. He saw how the displaced sand indicated that the animals were moving up or down steep slopes, 'cautiously sliding one paw downwards, till its footing became secure, and then extending the other in the same way, while its hinder feet, [were] following alternately ...' Duncan also suggested that the deeper front-foot prints of some tracks indicated the animals were bulkier around the head and shoulders.

The kind of animals that made these Permian imprints was not yet understood. The world was still reeling from the description of the first dinosaurs. The idea that anything mammalian – even in the loosest sense – had occupied an ancient Earth long before man, had only recently been tentatively suggested, based on the discovery of the mammal jaw in Stonesfield.

Geologists knew that the rocks of Corncockle belonged to the New Red Sandstone, so-named to contrast with the much more ancient *Old* Red Sandstone. Most naturalists believed that only reptiles existed in the time of the New Red Sandstone. Therefore these first footprints were likened to those of the closest living reptile analogues: crocodiles and tortoises.

Buckland happened to own several tortoises. After seeing the trackways and being given the suggestion of their makers by Duncan and another correspondent, he decided to carry out some experiments. Naturally for an eccentric like Buckland, these involved pastry.

'1st, I made a crocodile walk over soft pye-crust [sic]' he wrote in a letter to Duncan, 'and took impressions of his feet … [second] I made tortoises, of three distinct species, travel over pye-crust, and wet sand and soft-clay …'[7] Buckland's wife supplied the pie-crust and Buckland supplied the tortoises. Where the crocodiles came from is unclear, but as Buckland had a penchant for eating them he probably also had access to live ones. The results: the marks matched the tortoises. He concluded, 'though I cannot identify them with any of the living species … the form of the footstep of a modern tortoise corresponds sufficiently well … so I conceive your wild tortoises of the red sandstone age would move with more activity and speed … than my dull torpid prisoners.'

The anatomist Richard Owen assigned the genus name *Testudo* for Duncan's original footprint specimens (from the order of reptiles Testudina, which includes tortoises, turtles and terrapins), but the trackways were later rechristened *Chelichnus,* meaning *turtle tracks.*[*]

Of course the similarities between the Corncockle tracks and Buckland's pie-walking prisoners were purely coincidental. Tortoises and their kin didn't appear until the Triassic.[†] In the following decades after the initial description more prints were named from Corncockle and other Permian sites, including the Hopeman Sandstone Formation around Elgin. Authors suggested many different reptiles and amphibians as the track-makers, but it wasn't until the end of the nineteenth century and beginning of the twentieth that scientists began to associate some of them with the group most likely responsible for traversing the damp sands of deep time. They were closer kin to you and me than any reptile, and left their indelible marks not only on the rocks, but on the pages of our evolutionary story.

---

[*] *Testudo* is a genus of tortoises. To begin with there was no separation between scientific names for animals and for their footprints. Later, footprints were assigned to their own ichnogenus and ichnospecies, separate from the animals that are thought to have made them.
[†] The evolutionary origins of testudines and where they fit in the reptile family tree is in fact one of palaeontology's biggest mysteries. One thing is certain, there were no Permian tortoises.

To track these creatures down, we must first travel to Texas. For most people, the state conjures images of cowboys, cattle, conservatism and great hospitality. But Texas is red in more ways than one – the rocks beneath Texan boots are stained with the russet of iron-rich sedimentary rocks, and filled with fossil fuels and fossil synapsids.

Sedimentary rocks are formed from fine particles like sand and mud being deposited in layers, and over time petrifying. The 'red beds' of rock that form the foundation of much of Texas, as well as neighbouring New Mexico and Oklahoma, are some of the deepest deposits of Permian rocks in the world, at over 1,000 metres deep. Deposited in a warm delta teeming with fish and amphibians, the older rocks in particular record a wet, humid landscape that lay just south of the equator. These rocks are part of a sequence of layers (known as formations) covering around 86,000 square miles and 225 million years of geological time, spanning the Ordovician to the Late Permian. This is one of the few places on Earth that captures the fossil record of the earliest synapsids.

In the previous chapter, we met the first members of the synapsid lineage. Almost indistinguishable from their reptile-line cousins, they made an unpromising beginning to the story. As the Permian dawned however, the synapsids radiated out from their humble beginnings.

Four main groups emerged: the pin-headed caseids; the long-snouted ophiacodontids; the vegetarian edaphosaurids; and the carnivorous sphenacodontids. Together these animals are often called pelycosaurs, another unfortunate name as it means 'pelvis lizard'.* These animals were certainly not reptiles. The term pelycosaur is outdated, but still commonly used to refer to this group and distinguish them from the other synapsid groups that appeared later in the Permian. The name may feel unfamiliar to you now, but I'm betting that you already know at least one pelycosaur.

---

*The name pelycosaur is also problematic because the first part of it is ambiguous in Greek, meaning a wooden bowl, axe or basin. It originally referred to the shape of the ischia in the pelvis, which was a diagnostic feature of the group.

Many writers have covered the story of America's two infamous squabbling fossil scientists, Edward D. Cope and Othniel C. Marsh, and their all-consuming passion for dinosaur discovery. But in the course of their 'bone wars' they found much more than just giant reptiles. Cope was among the first white palaeontologists to explore the red beds of Texas, and in 1878 he described multiple fossil animals from there.[8] They were originally sent to him by Jacob Boll, a Swiss–American naturalist, but Indigenous Americans had already discovered these bones long before – for example, the Comanche knew them from sites in what is now Oklahoma.[9] Cope named many new kinds of early reptile, amphibian relatives, and even lungfish. But among his enduring legacies is the discovery of a character that remains one of the few early synapsids to sail to fame and retain instant public recognition to this day.

That animal is *Dimetrodon*.

The reason why *Dimetrodon* made it into the public consciousness when the other pelycosaurs didn't is somewhat of a mystery. It probably has something to do with the iconic outline it strikes. I often say to people that it is 'the one with the sail-back', and nine times out of ten they know which one I mean.

Dimetrodon has a sprawling posture and long body, superficially 'reptile-like'. Its big head is full of pointed teeth, and it has a tall skin-covered sail down its back. Many people's earliest encounters with it are from cartoons: in *Fantasia*, you may recognise it lolling by a pool enjoying Stravinsky. As in its cameo in *The Land Before Time*, *Dimetrodon* has somehow travelled over 200 million years into its future to hang out with pterosaurs and saunter past a group of lost baby dinosaurs.* There were, of course, no dinosaurs when *Dimetrodon* and co were making their home on Permian Earth.

Like an erroneous fin, *Dimetrodon* had a bizarre series of tall bone masts on its back, each one projecting skywards from the vertebrae. These projections are called neural spines, also known as spinous

---

* I love that film, but it's a hot mess when it comes to accuracy, not that I expect scientific rigour from my cartoons. The animals depicted all existed at totally different times from one another.

processes, and we all have them. Reach around and feel your own
backbone: can you feel the knobbly bits? Those are the neural spines on
each of your vertebrae. In humans they are quite small, but if you look
at them in other animal groups you'll find they vary in size, and often
vary in different parts of the spine. They are most obvious in the
shoulder area of browsing herbivores like cows and horses. Bison have
particularly impressive ones, forming a characteristic hump behind the
neck. The spines in those animals are for attachment of the muscles of
the neck. When you spend all day lowering and raising your head to
eat, you end up with a beefy neck and shoulders, and those big muscles
must attach somewhere. The neural spines provide this anchor.

The purpose of the neural spines in *Dimetrodon*, however, has been
a subject of heated debate for over a century. Their shape and
arrangement – shorter at the shoulder and hips, longest in the middle
of the spine – means they are almost certainly not for attaching muscles
used for feeding, or any specialised movement other than normal
activity.* The spines in the middle could reach two metres in height
on an animal that could grow over four metres long – that's much
bigger than the average three-seater sofa. What could this giant need
with such conspicuous accoutrements?

Grappling with the purpose of these elongated spines back in 1886,
Cope mused 'The utility is difficult to imagine. Unless the animal had
aquatic habits, and swam on its back, the crest or fin must have been
in the way of active movements.'[10] He noted the same strange
projections on other early synapsids. Some, like *Edaphosaurus*, not
only had the main vertical spines, but also secondary small spines that
projected horizontally from them at right angles.

> The animal must have presented an extraordinary appearance. Perhaps
> its dorsal armature resembled the branches of shrubs then, as they do
> now, and served to conceal them in a brushy or wooded region; or,
> more probably, the yardarms were connected by membrane with the
> neural spine or mast, thus serving the animal as a sail with which he
> navigated the waters of the Permian lakes.[11]

---

* At the end closest to the body, the spines have attachment scars for the muscles
of the back, but this is normal for all backboned animals.

Cope really took the idea of masts literally, believing the structure was more amphibious than terrestrial and comparing the animals to ships at sea. Using the spine bones as a wind-catching mast would have required pelycosaurs to surf sideways across bodies of water. It challenges all sense to think that this was a more useful evolutionary solution to getting around in water than just learning to swim like any other beast on the planet. We now know *Dimetrodon* was definitely a land-living animal. The 'sail' was nothing to do with swimming, or catching a breeze.

The most common interpretation you'll hear for the purpose of synapsid back spines is thermoregulation. This idea emerged in the 1940s, with heavy-hitters like Alfred Romer backing it as a good explanation. The theory was that *Dimetrodon* and fellow synapsids were ectothermic, or cold-blooded, like modern reptiles. To heat themselves up, they would have basked side-on to the sun and absorbed the warmth. The sharp teeth of most sail-backed synapsids meant they were predators, so heating themselves up allowed them to move more rapidly than their sluggish, non-sailed prey, giving them the advantage. If they got too hot, they could move into the shade and pump their blood into the sail, where it would cool them down more rapidly.

The use of sails for regulating temperature seemed to make intuitive sense. These structures even grew larger as the animal grew to bigger sizes – a relationship with body size is what you'd expect if heat exchange was the sail's purpose. Researchers had carefully examined the spines of *Dimetrodon* and *Edaphosaurus*, and even reconstructed the path of blood vessels (vasculature) in the bones. In 1986, a researcher called Steven Haack took things even further, using the mathematics of solar radiation, orientation of the sail, convective heat exchange and metabolic heat production to precisely calculate the performance of *Dimetrodon*'s back in thermoregulation. He concluded that the sail resulted in a 3–6° Celsius increase in body temperature during the day: 'the core temperature increases steadily over an interval beginning about an hour after sunrise and ending an hour or two before noon'. Just in time for lunch.

This small change in temperature was far less than calculated by other proponents of the thermoregulation hypothesis. 'The sail's effectiveness is not as impressive as one might have hoped', Haack admitted in his conclusions.[12] He also found that the sail was of very little use in releasing heat, so it was unlikely to help *Dimetrodon* cool down again. *Edaphosaurus*, on the other hand, seemed to be even less able to use the sail for raising temperature than its cousin. According to researchers a decade later,[13] those strange horizontal projections on *Edaphosaurus*'s spines that had so baffled Cope appeared, if anything, to make the sail better as a structure to cool the animal down rather than warm it up. This was because the cross-bars created turbulence in passing air currents, and provided more surface area for air to pass over and lose heat from the animal's blood.

These models and others like them were based on a lot of assumptions about the physiology of these animals. Notably, that they were ectothermic, or cold-blooded. Being cold-blooded would have meant they had no way to heat their own bodies other than using their environment. Modern ectotherms like lizards need to bask to raise their temperature, but of the thousands of species of cold-blooded sunbathers on Earth today, none have any kind of spiny back structure to assist them. What's more, back sails also occurred in small pelycosaurs, despite the fact that the physics of heat exchange changes with size. *Dimetrodon teutonis*, for example, stood at the height of a ruler (up to 30 centimetres or 12in), and at that size a sail would have little to no effect on heating or cooling the body. Examining the cross-section of the spines in microscopic detail, researcher Adam Huttenlocker and his colleagues also found little evidence that any major blood vessels ran through or around them. The holes that were previously given as evidence for increased blood flow were likely to be the result of rapid growth patterns, rather than vascularisation.[14] Another theory was needed to explain these extraordinary structures.

In the early 1900s it was proposed that pelycosaurs may have grown long neural spines as support structures for big fat humps. Humpback synapsids might have stored fuel in the form of fat deposits, clustered around their spines. *Edaphosaurus* seemed an especially likely candidate, as those extra horizontal struts could have provided structure for thick connective tissues which held the fatty hump and kept it stiff. Up to the 1970s some researchers were arguing for this interpretation. Unlike *Dimetrodon*, *Edaphosaurus*'s sail didn't get bigger as the animal's body size increased, which seemed to support the hypothesis.

However, the evidence against the fatty-back hypothesis began to stack up. Researchers examined some of the other animals with large neural spines. There were several examples of long spines that had developed in totally separate animal groups. Even *Dimetrodon* and *Edaphosaurus*, as similar as they were, were only distantly related, coming from two different groups of pelycosaurs. *Dimetrodon* was one of those meat-eating sphenacodontids, whereas *Edaphosaurus* gave its name to the herbivorous edaphosaurids. The spinal structures had evolved through convergent evolution, rather than being inherited from a common ancestor. Often, convergent evolution happens when different animals are trying to solve the same problems of survival, so it is logical to assume that these animals had back spines for the same reason.

Then there are the sail-backed dinosaurs, like *Spinosaurus* and *Ouranosaurus*. Could they have had fatty flesh-Mohicans running along their length? In 1998, a researcher named Jack Bowman Bailey argued that they did. In his paper on the subject, he pointed to the shape of dinosaur back spines and compared them to living animals, arguing that the dinosaurs' structures were similar to today's bison.[15] Furthermore, he pointed to the ossified tendons in the fossils of *Ouranosaurus* as evidence to support the idea. These tendons ran between the back spines and had turned into bone, making the structure even more rigid.

The humpbacked dinosaur theory was interesting, but it didn't help solve the mystery of the pelycosaurs. In comparing them, Bailey found dinosaur spines were not especially like the Permian animals' in their shape: the dinosaurs had very robust, oar-like spines, whereas the pelycosaurs' spines tapered like knitting needles. The pelycosaur spines were also much longer in relation to their body size, making up around 65 per cent of the animal's height, as opposed to less than 35 per cent in dinosaurs, and 45 per cent in bison.

There are also issues with the underlying interpretation of animal humps made by Bailey and other researchers. There are two living species of bison, *Bison bison* and *Bison bonasus*, and although they both have a handsome hump at their withers, these are mostly the result of having massive muscles, not storing fat deposits. Animals that do have fatty fuel-storage humps – camels famously have one or two on their backs – usually have no tell-tale hump-support structures in their skeleton. Neither do animals like the little fat-tailed dunnart, *Sminthopsis crassicaudata*, which stores fat in its tail. Tall neural spines, as I mentioned previously, are normally for supporting muscle, not cellulite.

Putting aside what dinosaurs were doing with their weird spine bones, it seems unlikely that *Dimetrodon* or *Edaphosaurus* were fat-backed. Some more radical ideas included the theory that the bones had no skin connecting them, but instead were a literal series of spikes jutting out in a defensive wheel. A fossil of *Dimetrodon giganhomogenes* from northern Texas showed evidence that the spines had been broken and had re-healed.[16] Researcher Elizabeth Rega and colleagues found the microstructure of those bones indicated that the spines were more resistant to breaking than the limb bones, supporting the idea that it might be attack-proof. But from what assailant? Being pretty high up on the food chain, it's unclear what *Dimetrodon* would have been deterring, except perhaps fellow sphenacodontids. The spines were very thin and tapering, and this makes them unlikely to provide any real defence. Animals that use spines for defence today are usually small-bodied, like hedgehogs and porcupines, and they tend to spread their spines around in an unappealing pin-cushion, not line them up like fence-posts.

Cope had suggested another hypothesis: camouflage. Perhaps the branches of *Edaphosaurus*'s spines made it look like a bush? Or *Dimetrodon*'s spikes concealed it among the reeds beside a river? Again, a look at living animals doesn't support this idea. Most ambush predators achieve cover through stealth and colouration. Animals with eccentric growths to provide camouflage are almost always small-bodied, and use it to hide from predators, not to become predators themselves. Although it's not impossible, the idea that these pelycosaur billboards actually helped them blend in seems unlikely.

There is only one explanation remaining: pelycosaurs were sexy beasts.

↓↓↓

Sexual selection is one of the strongest forces in natural selection, and has some of the most ridiculous results. Theories of sexual selection used to be framed entirely from a male perspective, with focus on what Charles Darwin described as 'a struggle between the males for possession of the females'.[17] With their outrageous embellishments, eye-catching plumage, improbable appendages, massive bulk and frankly silly dances, male members of the animal kingdom caught the eye of more than just their mates. Entire museum collections have been biased towards them, as research recently showed.

A study by Natalie Cooper and colleagues[18] discovered that in natural history museum collections, especially of birds, 60 per cent of specimens were of male members of the species. This was true even when the male members of species were less representative of the overall variation in a group. The authors suggested that one major cause of this bias was a 'deliberate selection for large, "impressive" male specimens, especially where males are larger or more colourful than females, or possess ornaments or weaponry such as horns or antlers.' This skew in collections has an impact on our understanding of animal diversity, for both the public and scientists. It also impacts research by creating a bias in sampling, and potentially overlooking important aspects of animal biology.

The original interpretation of the mechanisms of sexual selection was also gender biased. In many scenarios it is actually the female of the species that holds all the power, as they are the ones doing the choosing. This puts them in the driving seat of sexual selection, especially when it comes to the fantastically extravagant adornments of animals like birds of paradise.

One of the most compelling explanations for the pelycosaur sail-back is that it served to attract a mate. Like a peacock's tail, the size of the structure might have impressed the ladies, serving as a sign of fitness. It may have been brightly coloured, dazzling all and sundry. Sexual selection could even explain breakages. Seen side-on the large back fin would have made a pelycosaur appear much larger and more intimidating to rivals. Physical fighting is actually quite rare in the animal world, with most creatures doing their best to avoid coming to blows for fear of being injured. Watch those classic scenes in wildlife documentaries where male animals like ibex circle one another and size each other up. It happens in every size class and every vertebrate group – even invertebrates get in on it. From silverback gorillas to staghorn beetles, animals raise themselves, puff up and face off, hoping to scare their competitor. And if they are well matched … well, that could be how that Texan *Dimetrodon* acquired his fractures.

Of course, even this theory isn't perfect. Detractors argue that there is no clear difference between male and female pelycosaurs and the size of their neural spines. In most animals, a structure like this driven by sexual selection is larger in the male, and sometimes absent from the female altogether – hence the bias in museum collections highlighted by Cooper and her colleagues. This difference between males and females is called sexual dimorphism, and it can be caused by a number of different selective forces. In many insects, fish and reptiles, females are the larger sex, particularly if they have to invest in nest defence or survive food shortages. We are, as mammals, generally more familiar with the opposite pattern of having larger males, which is often (though not always) driven by sexual selection.

Theoretically, we should be able to use the size of the sail compared to body mass to separate male and female sail-backs into two distinct groups. But there simply aren't enough fossils to be certain if pelycosaurs were sexually dimorphic or not. Although it is generally true that display structures are larger in males, it is not always the case. Female *Dimetrodon* may have also needed to warn off rivals for food, territory and mates, minimising the physical differences.

Whatever the reason for having them, the tall sail-backs of pelycosaurs like *Dimetrodon* have helped them win at least one battle: the fight for an enduring place in the human consciousness.

From the time of the discovery of pelycosaurs, Western scientists realised that they were more like mammals than reptiles. 'Thus the reptiles and batrachia of the Permian period,' said Cope in 1880, 'resemble each other and the Mammalia more closely than do the corresponding existing forms.'[19] He and many others found similarities in the limbs and skulls of these animals that hinted immediately at their affinity – though it didn't stop people calling them 'mammal-like reptiles'.

One of the first major branches of pelycosaurs are caseasaurs. They included two families, the caseids and their closest kin, the eothyridids. Eothyridids are incredibly rare in the fossil record. There are just three genera at present (although shortly after writing this a new paper came out that tentatively suggested a fourth, *Asaphestera*, who we met in the previous chapter), and of those the namesake *Eothyris* is known from just a single skull. But what a skull: just six centimetres long, but wide and flat with a ferocious set of teeth. Most strikingly, this little animal had not one, but two sharp canine-like teeth on each side of the jaw, providing a double dose of nip in every bite.

The caseids' teeth and bodies tell a completely different story from the eothyridids. Named for their micro-headed representative, *Casea*, caseids had bulbous little heads filled with blunt teeth. They didn't sport sails, but had stocky long bodies with equally long tails. They had big noses – literally and in terms of nostril size – and wide stubby skulls. And they had ridiculously minuscule noggins.

To say *Casea* had a small head is like saying Antarctica is a bit chilly. Where the head might reach about 20 centimetres (8in), it was attached to a barrel of a torso and tail that reached 1.5 metres in length (4ft 9in). Its sister, *Cotylorhynchus*, grew to around 4 metres (over 13ft), but maintained similar pea-skull proportions.* Inside *Casea*'s mouth, rows of small teeth on the palate and a well-developed attachment for the tongue tell us this was a herbivore, using its equipment to pull up and crush plants against the roof of the mouth before sending them on their way. In this manner *Casea* and kin could chomp through the newly flourishing ranks of gymnosperm plants.

Caseids were among the first specialised herbivorous animals, but there were others. In the very latest Carboniferous there was the pelycosaur *Gordodon* (related to *Edaphosaurus*) and an animal called *Desmatodon*, from a disputed group of creatures recently argued to be synapsids, but possibly closer to reptiles. *Diadectes* is another

---

* Admittedly, having a small head was not confined to the caseids – *Edaphosaurus* was also possessed of a similar teeny cranium. For plant-eaters that don't need to do much chewing there's no need for a big head and mouth.

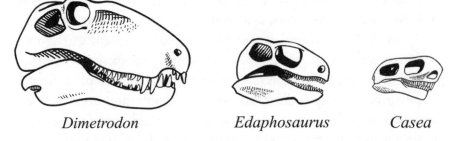

Dimetrodon   Edaphosaurus   Casea

Figure 2 Pelycosaur skulls varied from sharp-toothed carnivores (Dimetrodon) to pin-headed herbivores (Edaphosaurus and Casea).

example, an impressively chunky animal that belonged to its very own branch of tetrapods, probably amniotes. Whatever the family relationships of these beasts, it is clear that it was in the Permian that herbivory really took off.

This may not seem important, but developing a plant-based diet was a landmark moment for tetrapods. Insects and micro-organisms had already hit upon vegetarianism as a way of life, but it posed some unique challenges for vertebrates. One of the biggest problems is that plants are made of tough stuff. Their cells are mainly composed of cellulose, which can only be broken down with certain enzymes that vertebrate animals don't possess. This meant that despite the flourishing foliage, the animals of the Late Carboniferous and Early Permian couldn't get enough nutrition from eating them to make it worth their while. For the first tetrapods, it was like walking through a well-stocked supermarket with an empty wallet.

So how did caseids solve the problem? Working this one out was going to take guts.

To get around the problem of indigestible plant cells, herbivores harness the power of microbes. When we think of plant-eaters today our thoughts turn to farm animals like cows and sheep, or wild animals like wildebeest or llamas. All of these mammals are ungulates, a massive and wildly successful group of animals that have made eating veg their trademark move.

Ungulates are usually divided into two groups based on their toes: the odd-toed ungulates, or perissodactyls, and even-toed ungulates, or artiodactyls. This puts horses, tapirs and rhinos in the first group, and the rest in the other, and reflects a deep split between these lineages. The *number* of toes is not quite as clear-cut as the name suggests; rather it is which digit takes the body weight that really distinguishes the groups. In the perissodactyls weight is placed mainly on the third digit, the middle finger and toe. In artiodactyls, however, the axis of symmetry runs between the third and fourth digits, resulting in their characteristic cloven hoof. Feet, you might think, are not an especially fundamental difference between animals. But there is another deeper physiological difference between perissodactyls and artiodactyls, and that is in their stomachs.

Perissodactyls are hindgut fermenters, whereas artiodactyls are foregut fermenters. Along with elephants, rodents, rabbits and koalas, the perissodactyls solved the problem of plant-digestion by fermenting their meals in the lower intestine and caecum – the latter being a specialised pouch attached to the large intestine that holds a wealth of friendly, plant-eating bacteria. We have a caecum, but in most hindgut fermenters it is much larger and plays a more active role in digestion. Digesting by this method means hindgut fermenters can process food rapidly, survive on low-quality nutrition, and reach impressive sizes – some of the largest megaherbivores of the last 66 million years have relied on the hindgut. But this speed comes at a cost. These animals need a lot more food because they extract nutrition less efficiently than their foregut-fermenting relatives. This drives some of them to cover huge distances in search of sustenance. They often eat their own dung to get the full nutrition from their food (especially small-bodied hindgut fermenters). While that's not uncommon in the animal world, it's hardly a hobby to brag about by human standards.

Foregut fermenters have approached the problem differently. As well as artiodactyls, some marsupials and rodents, sloths and the hoatzin (a type of bird) have convergently hit upon foregut fermentation as a way to get nutrition from plants. This method of digestion has resulted in some incredible changes to their biology.

Foregut fermenters include ruminants, such as cattle, sheep, goats, deer and giraffes, which have developed multi-chambered stomachs. Non-ruminant foregut fermenters, such as marsupials and sloths, have an enlarged and elongated stomach without multiple chambers.

Each chamber in a ruminants' stomachs is dedicated to a different stage in the process of digestion, aided by a host of friendly bacteria. The first two chambers are the rumen and reticulum. They comprise the largest portion of the stomach, and this is where the bacteria really get to work breaking down plant cells. Food from these chambers is usually brought back up into the mouth for a second grinding, also known as 'chewing the cud'. If you're inclined to get that close to one, you'll notice that foregut fermenters often have sweet-smelling vegetable breath, the result of repeatedly chewing on their fermenting gut-contents.

Food passes through the next chamber, the omasum, before finally finding its way to the 'true' stomach, or abomasum. In there, any friendly bacteria left in the cud meet their untimely demise, as gastric juices break everything down in preparation for the journey through the intestines. It is from their abomasum that we get rennet, an enzyme secreted by the stomach of ruminants that humans use to make cheese. It's enough to turn your stomach. This four-stage process means foregut fermenters get a lot more nutrition from their food, and can survive on much less, but their longer digestion time means they can't bulk-process food. This difference in herbivore stomachs is why horses can survive in parts of North America where forage is too poor for cattle. It also places an upper limit on the body size of foregut fermenters.[20]

Whatever method you employ as a herbivore, eating plants requires the help of a host of micro-organisms, and you need somewhere to keep them. Plant-eaters characteristically have big bodies filled with fermentation chambers and lots of intestine, to squeeze all the nutrients they can from their food. Those that don't have to chew can manage meal-times with proportionally smaller mouths than meat-eaters – it is, after all, just a hole to put the plants into. As we discovered from looking at neural spines, big neck and shoulder muscles are helpful, and blunt, spatula-shaped teeth clamp down on leaves and shear them off from their stalks.

The rules have changed very little in 280 million years. The broad deep body of caseids like *Casea* and *Cotylorhynchus*, and edaphosaurids like *Edaphosaurus*, probably held a fermenting chamber teeming with bacteria, ready to help extract every last bit of goodness from the plants they ate. How they acquired their bacterial helpers is uncertain, and fossils are unlikely to provide an answer. It has been suggested that micro-organisms may have initially been ingested by early tetrapods when they ate some decomposing plant matter, or plant-feeding insects. Eventually, some of the plant-processing bacteria survived in the gut, and a symbiotic relationship developed with the host, as animals with more micro-organisms extracted more nutrients from plant-matter, improving their survival.

Although herbivory evolved independently in multiple terrestrial animal groups, it is in synapsids that we find some of the first dedicated plant-eating adaptations, right back into the Late Carboniferous and Early Permian. Their lineage would continue to thrive at the forefront of this niche for the next 50 million years.

But the rocks of Clashach Quarry are much younger than this. They date from near the end of the Permian, by which point a whole new wave of dynamic beasts had replaced the pelycosaurs of old. It was they who would beta-test the modern food chain, and produce survivors that taught us major lessons about extinction and recovery. Theirs was the peak of the first age of mammals.

Moschops

Bos

Estemmenosuchus

Oudenodon

Biarmosuchus

Gorgonops

Canis

Glanosuchus

Tiarajudens

Suminia

Felis

Procynosuchus

Diictodon

CHAPTER FIVE

Hot-blooded Hunters

… to the primeval mammals their great tusks and their sharp claws, – that he of old divided all his creatures, as now, into animals of prey and the animals preyed upon …

Hugh Miller, *Testimony of the Rocks*

Fossils are like echoes in the caves of geological time. Like echoes, fossils can deceive you. What you hear as you reach out in the darkness is distorted; small sounds become thunder, what is behind you is in front. Those that stay silent and still are never found. Loud shouts from other chambers are swallowed by the rock.

As palaeontologists, we are often groping through the fossil record trying to find our way. When the record is good, it's as if someone switched on floodlights, and we map the stalactites and photograph the paintings on the walls. When the record is bad, we stumble blindly.

Studying footprints, it rarely feels as though the lights are on. To the casual observer, they seem quite plain to see and understand. But finding fossil footprints is not like looking at modern ones because we almost never know what animal actually made them. If you find an animal track today you can look up the fauna of that region, providing you with a list of suspects to narrow down. But two crucial pieces of information are often incomplete or missing for the ichnologist (those who study trace fossils like footprints): how old are the rocks, and what was alive at the time?

The most obvious way to know how old your rock is, is by observing how it relates to other rocks around the world. Logically, the rocks on top are younger than the rocks at the bottom. Geology plays games with us however, because it can lift and flip the layers like pancakes. There are times when rock is not being deposited at all, or layers are removed by processes of erosion – like the action of ice scraping across the landscape, removing pages from the book of time. This has frequently confused researchers. In the quarries around Elgin in Scotland, for example, rocks from the Permian lie directly on top of rocks from the Devonian, missing the Carboniferous out completely.

The biological contents of rocks provide the next clue to age. As we have come to understand more about the patterns of evolution, we have identified where groups of animals and plants appeared and disappeared. This is called faunal succession. By noting which groups are present in the rock, scientists can narrow down where in the sequence that rock must sit. Species of ammonite and trilobite, iconic marine fossils that were very speciose, as well as plant pollens and seeds, have proven especially useful for rock dating.

It was faunal succession that alerted Victorian scientists to what was happening in the rocks around Elgin. They found fish in the deeper layer – ancient armour-plated strangers from the age of fishes. In the rocks above, however, were the bones and footprints of fully land-living vertebrates. Clearly, those upper layers were much younger than had previously been supposed.

In the last 100 years, geologists have added radiometric dating to their toolkit. The result of combining faunal succession, radiometric dating and other methods to determine the age of rocks, is the chronostratigraphic chart. In common parlance we often call it a geological timescale. This is the familiar diagram you'll probably have seen in books and television series, with the names of all the different units of time stacked up in order, and colour coded with their dates alongside. An official version of this chart is released by the International Commission on Stratigraphy every few years, which amalgamates the latest and most accurate information we have on how old our rocks are (you can download it for free at www. stratigraphy.org).

Running an eye down the chronostratigraphic chart, you may notice that as we head back through time, the dates are often shown with another number beside them and a symbol that looks like a plus straddling a minus sign: ± . That is exactly what this symbol means: the date is plus or minus that number. These are the error margins. Even the most precise dating of rocks has one, and they can vary from hundreds, to thousands, to millions of years.

Most people are familiar with carbon-dating, which uses the natural decay rate of carbon to calculate the age of biological samples. This method works because approximately every 5,700 years the amount of a particular type of unstable carbon isotope, Carbon-14 (^{14}C), halves. This is called its half-life.

During your lifetime you absorb ^{14}C from your food, and your level matches the world around you. But when you die and stop replenishing your ^{14}C, it begins to decay and turn into the much more stable form of carbon, ^{12}C. To date a sample, scientists measure how much ^{14}C there is compared to ^{12}C, and they compare that to the level in the atmosphere. Because they know how fast the unstable carbon decays, they can calculate how long it was since that sample stopped replenishing its ^{14}C. That is carbon-dating.

But beyond around 50,000 years ago, carbon-dating becomes unreliable, and pretty soon it can't help you at all. This is because the amount of ^{14}C becomes too small to measure. To work out dates older than this, scientists turn to other elements like potassium and uranium. Similar to the way carbon decays, a particular form of potassium (^{40}K) decays to become argon (^{40}A), with a half-life of about 1.2 billion years. Uranium (Ur) decays via more than one route, with a half-life of 710 million years or 4.47 billion years, depending on which route it takes. Either way, it becomes lead (Pb).

The time-scales are brain-boggling, but by calculating changes in these elements it is possible to work out approximate dates for the ages of rocks. Complications in the calculations mean that the results always provide a range, not an exact pinpoint date. These error margins are nothing to fear; they are a sign that the science is being done carefully, and provide a measure of how precise the answer is.

But not all rocks give good dates. The best candidates are volcanic rocks, because we know that they erupt and rapidly cool on the surface of the Earth, capturing their chemical information like an original photograph. Sedimentary rocks, on the other hand, are more like a collage. The small particles that they contain could be many millions of years older than the rock itself; they might have been washed off the side of a mountain and mixed with organic matter to make a composite. This jumble of time defies most chemical-dating techniques. The compensation for this is that it is in sedimentary rocks, rather than volcanic ones, that we find fossils. Palaeontologists must use the more precise dates of volcanic rocks above and below to narrow down the time period of their fossils. They then compare their fossils with those of the rest of the world, and our knowledge of evolution and faunal succession. This constrains the timing further, making it possible to place extinct life within the framework of geological time.

But when palaeontologists say they've narrowed down the timeframe, it may be wider than you were expecting. Working habitually in millions of years gives those who study fossils and rocks an odd perspective on what is recent or ancient. To scientists working on the fossils of the Permian, even dinosaurs seem a bit too 'modern' for their tastes. Whatever time period you focus on, you'll find that to a palaeontologist the kind of error margins presented by radiometric dating are nothing to worry about. It is rare that you can take a photograph of the fossil record that is more than a few million years in resolution.

For the ichnologist working out who might be responsible for fossil footprints, the error margins of geological time provide a lengthy suspect list. This can be winnowed down by the careful evaluation of the shape of the footprint. Ideally, footprints should reveal foot shape, and so Cinderella marries the prince. Of course in real life billions of people wear the same shoe size.

What's more, footprints are rarely high-fidelity. Like William Buckland trying to figure out who walked across the sand dunes of ancient Scotland, we can easily be misled. Sand shifts. Mud smudges. Feet slide and scuff, sink to form messy pot-holes, or barely impress on the hard ground at all. A sloping surface changes how the track-maker walks, rearranging their feet. Then there are the processes that come after the animal has passed by: rain fills the marks in, wind picks at the edges, flood blurs their impressions. Millions of years tilt the prints, chop them up and file them down. Unpicking these processes is often complex, and sometimes all we know is that some kind of animal long ago passed us by.

The footprints filling the walls of Scotland's Permian quarries are not the only examples of their kind. Similar prints are found in Germany, North America and Argentina. Recent research suggests they belong to animals from at least five different groups, including members of the next great radiation of synapsids to make the Permian their home.

From among the ranks of the pelycosaurs, a new group emerged. They would come to develop the key traits we associate with mammals, including warmer blood, higher energy lifestyles, and perhaps even fur. What's more, they established for the first time an ecosystem we still recognise today, one based on large numbers of herbivores fed upon by a smaller cohort of carnivores.

This revolutionary group are called the therapsids. It was these creatures that hiked across the hot heartland of Scotland, and beat a path that all mammals were to follow.

Almost as soon as they emerged from among their predecessors, therapsids were something special. From their first appearance in the fossil record in Russia, South Africa and China – the best places in the world to find their bones – these animals were diverse and radically different from everything else on Earth.

It's clear that the therapsids evolved from a common ancestor, forming a distinct group that share easily identifiable skeletal features. Their teeth had begun to differentiate, with canines and incisors at the front that were unlike the 'postcanine' teeth behind them. The bones of the face began to shift to accommodate the specialised canines. As we will see later in the mammal story, increasingly complex teeth dedicated to different food-processing roles are among the most important evolutionary wizardry practised by the mammal lineage.

The lower jaws of early synapsids were composed of multiple bones, a feature inherited from the first vertebrates. But those bones were changing. The largest one was the dentary – the one now comprising all mammal lower jaws – and this was steadily becoming the largest of the jaw bones, holding all the bottom teeth. The rest of the bones were at the back of the jaw. On one of these bones, called the angular, first a notch and then a sheet of bone appeared, called the reflected lamina. The function of this is not clear, but it may have played a role in the formation of the mammalian ear – another mammal superpower that emerges somewhat later in our story. The whole back of the skull became more robust, and the jaw joint stronger, while the fenestra that had long defined the synapsid skull enlarged. Taken together, these changes tell us that therapsids had developed bigger muscles around their skulls, making them better biters. This opened up a new culinary world.

In all, researchers have identified as many as 48 synapomorphies, special characteristics of the skeleton, that define therapsids. By comparing these features with the pelycosaurs that preceded them, it is clear that therapsids emerged from among the carnivorous sphenacodontids. They share so many traits that this relationship is as close to certainty as palaeontology gets. It means that the iconic sail-back *Dimetrodon* is actually among our closer, albeit distant, kin.

Although the changes seen in the skull of therapsids are important, they are far from the most radical alterations natural selection picked out. For the first time, our ancient mammal ancestors were beginning to *move* in ways we would recognise today. The earliest synapsids still had a sprawling gait similar to the one retained by many reptiles. The shape of their limb bones and how they fitted together limited the range of their movement – you couldn't high-five a pelycosaur. In therapsids all that changed. The shoulder bones reduced in size and they lost the complex interconnecting bones around their chests that had been a straitjacket to their ancestors. The hips altered too, and the largest bone in the leg, the femur, developed a rounded head that fitted snugly inside the hip-bone socket, rather than simply resting against it.

These sleeker, more streamlined arrangements meant limbs could be brought closer underneath the body. The implications for Earth's ecosystem would echo into the present day. To discover more, we must search for the bones of these animals in the vast country of Russia.

Before the Trans-Siberian railway was built, it is said to have been faster to sail over the Atlantic Ocean, cross America, and sail over the Pacific to reach the Russian Far East, rather than attempt the trek through Siberia. The railway is considered one of the greatest journeys on Earth. In my twenties, fuelled by a childhood intoxication with travel and *Dr Zhivago*, I spent two weeks experiencing the 9,289 kilometres (5,771 miles) from Moscow to Vladivostok across Earth's largest, starkest continent.

The journey was transformative, but I felt every mile. My shared cabin was freezing at night, with ice coating the inside of the cracked window. But it was stupefyingly hot and stuffy during the day. To escape I spent much of the journey standing in the partition between the carriages. It smelt of diesel and remained below minus 10°, but in my solitary, frozen world, I watched the sparse landscape and far-off mountains trundle past, humming 'Lara's Theme' or singing David Bowie's 'Life on Mars' quietly to myself like a lullaby.[*]

There was no Russia in the Permian. It's hard to image that those thousands of kilometres of taiga and mountains, about as distant from the ocean as it is possible to get in the world, have not always existed. Where the magnificent Ural Mountains now divide the great Asian continent in two, there once flowed a shallow sea called the Palaeo-Uralian Ocean. It separated the land masses that are now Europe and the westernmost part of Russia from Siberia, Kazakhstan and the rest of central Asia. As the Carboniferous drew to a close and the Permian wore on, this ocean closed, and the land beneath it crumpled. The folds rose to form the Urals, a 250 million-year-old healed scar. Russia is now part of the largest continent on Earth, but in the Permian it was merely the head and shoulders of a far mightier Pangaea.

When Scottish geologist Roderick Impey Murchison travelled across Russia and mapped the Urals around Perm in 1841, there was no railway to speed his journey – the first tracks weren't laid until 1891. It took 26 years and immeasurable human suffering and sacrifice to create the Trans-Siberian railway. It was hacked out of the earth by Chinese labourers, soldiers, exiles and convicts. Bodies lie beneath its surface: workers who died from floods, landslides, bubonic plague, extreme cold, cholera, anthrax, bandits and tigers.

There are far older bodies too, sacrificed as the bedrock of Russia itself was assembled. Russia is one of the best places in the world to find Permian therapsid fossils. Whereas earlier synapsids appear to have been restricted to the warmer, humid equator of Pangaea, their

---

[*] 'Lara's Theme' is the iconic composition from *Dr Zhivago* (1965), which runs through the film, and features the Russian stringed instrument, the balalaika. And Bowie used the Trans-Siberian to get back to Europe after his tour of Japan 1973. I wonder what he sang to himself along the way?

descendants were swiftly found all around the world, in both Laurasia and Gondwana.

These are not the cold-blooded 'mammal-like reptiles' of old parlance. They are the core of the first complex ecosystems on land, including swift hunters and dangerous browsers, climbers in the highest branches and diggers of deep soil. The height of the Permian was a Dreamtime,* when the Earth foreshadowed its later evolutionary plotlines.

Legendary animals like these require legendary names. One of the oldest groups, and most distant to our own branch of the tree, are the biarmosuchians. Biarmosuchia and its namesake *Biarmosuchus* are named from the Old Norse for the area of Russia around the White Sea, which they called Bjarmaland. The biarmosuchians had distinct large canine teeth, and something unseen in modern mammals: a ring of bones in the eye called a sclerotic ring. These support the eye, and their presence in biarmosuchians as well as some other therapsids and earlier synapsids (and in reptiles) makes it likely that they were common in their shared ancestors, and later lost.

Looking at the skeleton of *Biarmosuchus*,[†] it had an eager stance: it reminds me of a bull terrier ready to chase a stick. It was a bit smaller overall, but the head has the same strong profile. In the flesh, it wouldn't have had external ears – those skin flaps that stick out and pivot to hone sound received by modern mammals – only holes, leading to a simple internal ear, like those of their ancestors. And despite its forelimbs being a little more splayed out to the side, you have no trouble imagining it racing across the emerging landscape of Permian Russia, chasing dinner rather than a tossed branch. The pelycosaurs were dynamic, but *Biarmosuchus* and kin are undeniably built to move further and faster.

---

* The Dreamtime is a term commonly used to refer to the culture and spirituality of Aboriginal Australians, and is also known as 'the dreaming'. It often refers to the creation of the landscape and animals that populate it.

† The fossils of *Biarmosuchus* are actually thought to be a juvenile skeleton of what would have become a much bigger adult, but I've retained its description here for the sake of example. It might actually belong to *Eotitanosuchus* or *Ivantosaurus*. If so, the adult size could have been as much as five metres (16ft) long.

It is true that there were no dinosaurs in the Permian, but there were dinocephalians. The name has nothing to do with those distant reptiles, other than the shared root of the word: these synapsids were not 'terrible lizards', but they did have 'terrible heads'.* This is in reference to their intimidating faces, with a tendency to have thickened bones in the skull (called pachyostosis) – but don't let that fool you; dinocephalians were not all ferocious beasts. The group was one of the earliest to emerge, but their members included carnivores, omnivores and specialised herbivores. For a time they became the most numerous therapsids, and their bones can be found not only in Russia, but also in China, South Africa, Zimbabwe and Brazil.

One of the most charismatic dinocephalians is *Moschops* from South Africa. This animal was a serious 'chonky boi': longer than a super king-sized bed, and built like a nightclub bouncer. That bulk held the digestive chambers necessary for processing vegetable matter, because like most of the largest animals alive today, *Moschops* was a plant-eater. It had massively stocky shoulders and a short neck, and shared those defining features of dinocephalians, a thick skull, a steeply downward-sloping face and an equally steeply sloping back. All in all, *Moschops* looked like a triangle with legs.

In 1983 *Moschops* found television fame in the United Kingdom in a stop-motion cartoon series of the same name. It did well enough to make it across the Channel to Denmark. Like *Dimetrodon* however, this synapsid found itself misplaced in geological time in the cartoon, hanging out with a motley ensemble of Jurassic and Cretaceous family and friends, including an *Allosaurus*, Grandfather Diplodocus, Mrs Kerry the *Triceratops*, Uncle Rex and Mr Ichthyosaurus. But who can blame the cartoon makers; *Moschops* had a face for television, endearing enough to have earned it its name, which means 'calf-face'.

Moschops's brother *Estemmenosuchus*, on the other hand, had a face for radio. It was an omnivore from Russia that only a mother could love,

---

* Just to confuse things, there is even a dinocephalian called *Dinosaurus*, named a few years after Owen coined the term dinosaur. Dinocephalians are also not to be confused with *Dinocephalus*, which is a type of longhorn beetle. Presumably, their heads are also pretty terrible.

taking the thickening of the skull to extremes. Horns erupted from *Estemmenosuchus*'s head in all directions, as if someone had set off a firework in its skull. Two gnarly protrusions sat above the small rounded eyes, and another two exploded from the cheeks. Like *Moschops*, it was a real beefcake, reaching three metres (almost 10ft) in length.

A cast of *Estemmenosuchus*'s skull is on exhibit in the Royal Tyrrell Museum in Drumheller, Canada (the original, along with all the rest of the fossils known from this animal, is in the Palaentological Institute in Moscow). The skull in Canada is eerily lit with its mouth open in surprise – it never fails to elicit a similar response in the viewer. I felt a twinge of sadness when I visited the museum, watching a queue of visitors file past and photograph *Estemmenosuchus* for its grotesquery, doubly insulting the poor beast by assuming it was some kind of dinosaur. What may seem bizarre from our perspective, made this giant a force to be reckoned with by its contemporaries, and was perhaps also a measure of fitness used in courtship. Similar to *Triceratops* and other ceratopsian dinosaur copycats many millions of years later, *Estemmenosuchus* and fellow dinocephalians probably developed head adornments and pachyostosis for defence and competition, possibly as reinforcement for head-butting.[1]

It looks as if those therapsids had evolved yet another of nature's great innovations, and perfected it long before the rest of the world caught on.* These trend-setting critters were also the first to evolve some of the most iconic equipment possessed by any fossil animal: sabre-teeth.

One of the world's top favourite extinct animals of all time has got to be the sabre-toothed cat. Dinosaurs aside, this is such a recognisable animal, with no living equivalent, that it has become the multimedia star of books and films, a recognisable face in the pantheon of kids' beloved megabeasts. The fact that humans lived alongside it makes it even more compelling. But what we call sabre-toothed cats include a number of rather distantly related animals that have independently

---

* A slightly younger (more recent) Permian group of reptiles, the pareiasaurs, also developed head adornment shortly after the dinocephalians became extinct. Like them, they were big hulking herbivores.

grown excessively large canine teeth over the last 42 million years, through convergent evolution.

The most famous, and one of the most recent, sabre-toothed cats was *Smilodon*. It is also known by some as a sabre-toothed tiger, but although tigers and *Smilodon* are both cats (Felidae), *Smilodon* belonged to an extinct group separate from tigers. *Smilodon* is incredibly well known thanks to the hundreds of fossils pulled from the ooze of the La Brea tar pits in Los Angeles in the United States. It was a heavily built ambush predator, with such enormously disproportionate canines that it had to evolve an exceptionally wide mouth-gape to be able to fit anything in there to bite on.

There were, however, many other sabre-toothed 'cats'. In the same family as *Smilodon* was the closely related *Megantereon*, and a whole host of slightly more distantly related animals called nimravids, non-felid machairodontids and barbourofelids (collectively known as the 'false sabre-tooths'). And that's just the felids (the cat family) and their close sisters. Before the carnivorans[*] claimed their primary role as meat-eaters in the early Cenozoic, there was another flesh-specialising radiation of mammals called the creodonts. They developed their own sabre-toothed specialist, *Machaeroides*, which was feeding on the earliest horses over 40 million years ago. And it wasn't just in the northern hemisphere that such sabre-toothed hunters emerged. Marsupials, not to be left out of the fun, had their own unique take on this adaptation: *Thylacosmilus*, which emerged from an enigmatic group called the sparassodonts in South America around 21 million years ago.[†]

Although they differ in the exact shape and length of their canines, these animals prove that if it's a good way to make a living, a feature like this has a tendency to emerge again and again as a solution for

---

[*] The family Carnivora (plural, carnivorans) is a distinct group that includes animals like dogs, cats, bears, hyenas, weasels, mongooses, badgers, racoons, seals and sea lions. They possess specialised carnassial back teeth, which evolved to slice through flesh. Although most of them predominantly eat meat, it is important to note that not all animals in the world that are carnivor*ous* are carnivor*ans*.

[†] These are technically not marsupials, but belong to the larger grouping called metatherians (which include marsupials). We'll talk more about this distinction in later chapters.

survival. Enlarged canine teeth can be used not only for killing, but also for competition within species. If we include tusks, we see musk deer and walrus have them, as do camels, primates and most pigs. Tusks are usually elongated canine teeth (except elephants' tusks, which are incisors), but there is a subtle difference between tusks and sabre-teeth. Tusks grow continuously throughout life, whereas teeth are subject to replacement. This distinction can be tricky to apply to the fossil record though, where there are instances of what appear to be sabre-teeth and tusks peppered through the same family trees. Often there are not enough fossils from the same species at different life stages to tell one structure from the other.

Casting our eye back across the depths of mammal evolution, enlarged canine teeth are far from a new idea. Over 252 million years ago, therapsids include the first sabre-tooths. The two main groups to discover the benefits of enlarged canine teeth were the gorgonopsians and the anomodonts. Both would play fundamental roles in the story of mammal evolution, but both are now extinct, leaving no survivors.

No name is more legendary among therapsids than that of *Gorgonops*. It was christened by Richard Owen for the Greek Gorgons: three mythical sisters with hair made of living snakes, who turned people to stone with their eyes.* *Gorgonops* itself has long petrified, but it still provides a fearsome countenance. The animal was a predator that could grow to the size of a motorbike. It had long powerful limbs, suggesting fast movement, but the most tell-tale character of all regarding its lifestyle was its teeth: elongate canines that jutted well below the rest of the tooth row.

Gorgonopsians emerged late in the Permian, but swiftly became the main predators in the ecosystems of what is now Africa, Russia and India. They ranged in size from only around a metre in length (three feet) to bigger than a bear. Compared to some of the other therapsids, *Gorgonops* and kin were not particularly diverse. They retained some features from their ancestors that initially made them appear less derived than other therapsids, suggesting they belonged to a much earlier branch of the tree. It wasn't until researchers like Denise

---

* If you've ever seen a photo of Richard Owen, it looks like he could have done the same.

Sigogneau – later Sigogneau-Russell – turned their attention to them in the 1960s to 1980s that the relationships of this iconic group became clearer. Sigogneau-Russell, whose PhD thesis in 1969 was based on her work on therapsids in South Africa, would later become one of the most prolific researchers on European Mesozoic mammals, carrying out foundational research particularly on fossils from her native France and the British Isles.

Most of the attention lavished on gorgonopsians is focused on those oversized teeth. As well as pointy fangs, which were sometimes serrated along their back edge, the group had well-developed incisors at the front of the mouth that fitted closely together. Their back teeth, on the other hand, were severely reduced, and in a few species had been lost altogether. Meanwhile at the back of the skull, a large flange of bone developed to attach strong muscles for a powerful bite.

A major question has plagued the palaeontologists who study sabre-toothed animals. Just like *Smilodon*, we see in animals like *Gorgonops* a suite of related features of the skeleton that made it a top predator, perfectly suited for flesh-eating. But how do you go about biting and killing your prey with such enormous gnashers?

We can see from the jaw joint that these animals had huge gapes to allow them to sink their teeth into their quarry, but then what? You have up to a tonne of furious vegetarian bucking around in your mouth. Those daggers begin to look terribly fragile up against that kind of bulk, and a broken canine could seriously impede or even kill a carnivore if it had an impact on hunting success. Some authors have suggested the teeth were such a risk for hunting that killing may not have been the selective force that drove their elongation. For example, researcher Marcela Randau and colleagues examined the relationship between tooth size and body size in different sabre-toothed cats, and suggested that sexual selection may have driven increased tooth size, rather than meat-eating.[2]

Understanding sabre-tooth killing mechanisms has preoccupied palaeontologists. In modern carnivores the canine teeth are used to kill prey and to fight. But it is clear from the length and slender shape of sabre-toothed cat canines that they wouldn't have been able to subdue struggling prey without breaking their teeth. Another theory is that sabre-toothed hunters stabbed their food to death, using their teeth like a pair of kitchen knives to cause deep fatal wounds. This seems plausible at first glance, and the reinforced base of sabre-teeth supports a piercing action for the tooth rather than a slicing one. But there would have been too great a risk of striking bone and shattering a tooth during frantic stabs. It is also not clear how such a motion could be achieved without the lower jaw getting in the way.

For a time, the favoured hypothesis for *Smilodon*'s killer move was the shear-bite. The suggestion was that after bringing down prey sans biting, these animals would have inflicted gaping wounds on the softer exposed belly, then retreated to let the animal weaken and collapse. The finishing move would be a death bite to the throat; pinning the exhausted victim down with powerful forelimbs to prevent them from wiggling dangerously as their windpipe was crushed closed.

The jury is still out about sabre-tooth killing mechanisms. One thing is clear though: there is not a one-size-fits-all answer.

For gorgonopsians, unresolved arguments continue about how these first sabre-teeth were employed. Unlike the mammal groups of the last 42 million years, these therapsids lacked specialised back teeth to slice meat up before swallowing it. It has been suggested that the combination of powerful jaw muscles, long canines, a reinforced skull, and a lack of cutting back teeth mean that gorgonopsians probably ambushed their prey swiftly, causing massive damage to their fleshiest parts by inflicting deep bites and tearing wounds. Unlike sabre-toothed cats, gorgonopsians didn't have to worry about breaking their teeth during an attack – indeed, they often did – as they continually replaced their teeth, so a broken fang was soon replenished by a fresh one.* They may have then waited, as sabre-toothed cats are suggested to have done, until their weakened quarry was easier to deal with. The large incisors could be used to shear off chunks of flesh, which were swallowed whole.

Hunters and prey, shown in epic chase, are one of the most common depictions of nature in wildlife documentaries. We are enamoured of

---

* We don't know how long it took for this replacement to occur. Continuous tooth replacement is a feature lost later in the mammal lineage, as we will find out in coming chapters.

tales of fearsome predators, and instantly recognise this tango of survival danced again and again across the globe.

A simple food chain like this, with plant-eaters at its base and meat-eaters feeding upon them, is understood by all ages in all cultures. This framework for the natural world was first assembled in the Permian. However strange and ancient these animals were, we see in them the seedlings of the ecosystems that surround us today.

Anomodonts, our second group of sabre-toothed therapsids, look like vampire turtles that have crawled out of their shells. The later ones in particular were something unique to behold: their bodies were long and low-slung, their heads short and boxy, they had stubby little legs and tails. Not only did they join gorgonopsians in having some of the first sabre-teeth (or as noted earlier, tusks), but they coupled the teeth with something even more radical: a beak.

Teeth and beaks are not bedfellows in the modern world; we might even consider them mutually exclusive. After their first appearance in the fossil record in South Africa, China and Russia, anomodonts steadily reduced their teeth at the front of the face and replaced them with a beak, but they never entirely lost them. Their canines remained intact and grew much longer. The most extreme example is in the *Tiarajudens*, perhaps the first sabre-toothed herbivore to have existed.

Tiarajudens comes from Brazil. It was about the size of a pig, but sported dagger-like canines longer than your hand[3] – the exact length is not yet known, because the only fossil described so far is missing the tip of the tooth. The function of these teeth is certainly not killing; although *Tiarajudens* is an early anomodont and so lacks a beak, its mouthful of teeth are clearly shaped for nipping off and grinding vegetation. The large canines are almost certainly for display, and perhaps a squabble or two.[4] Although later anomodonts don't have the extreme sabre-like canines of their predecessors, they continue to wield prominent canines. These are usually referred to as tusks, and likely shared a similar purpose to those of tusked mammals we see in the modern world.

Meanwhile in Russia, another early anomodont was not to be outshone when it came to innovation. *Suminia* was about the size of a lemur. It is known from multiple skeletons, including 15 juveniles washed into a muddy flood plain west of the newly forming Ural Mountains. This particular anomodont had a body plan modified for a way of life no other synapsid had grasped before. It was arboreal: the first tree-dweller of the mammal lineage.*

Living in trees presents a series of physical challenges. First of all, you have to have a small enough body to be able to get yourself up there. Most small creatures can clamber up a tree, but that doesn't make them arboreal. To actually live in the treetops requires dexterity and grasping hands and feet. *Suminia*'s skeleton suggests it was the first tetrapod to develop these traits. Long fingers and toes, with proportions more like modern tree-dwellers than its closely related anomodont siblings, suggest it clung to branches.[5] Long limbs also support this interpretation. It may even have had a prehensile tail, able to curl and grip to provide extra help to stay aloft.

So far, no other arboreal anomodonts have been found, but if *Suminia* really was a specialist tree-dweller, it adds yet another distinction to the therapsid CV. The Permian may have been ancient to us, but it represented a diverse, modern ecosystem.

By the Late Permian, just over 252 million years ago, one group of anomodonts called dicynodonts had lost their front teeth entirely. The beak that grew to replace them was tough and keratinous like a turtle's, and attached to powerful musculature. These animals could clearly snip through even the toughest stalks. It was to prove an excellent combination, and soon dicynodonts ranged from rabbit-sized burrowers to creatures of hippo stature. What's more, this group played a significant role in the epic tale of mammal evolution because they proved to be consummate survivors in the face of disaster, as we will find out in a moment.

And so as the Permian drew to a close, Pangaea was awash with therapsids. They had developed a huge range of body sizes and eaten

---

* Although there might be an even older contender. An animal called *Ascendonanus* from the Early Permian is thought to be arboreal. It belongs to a group called varanopids, which were once thought to be synapsids. As I wrote this book however, a new paper came out by my colleagues at the University of Oxford arguing that they were actually more closely related to reptiles.[6] This is the trouble with writing a book about science, it never stands still. Who knows what else will have been discovered by the time this book is published?

their way through the landscape – and one another. The first recognisable modern food chain had emerged, with caseids, biarmosuchians, dinocephalians and finally dicynodonts taking advantage of the smorgasbord of greenery, accompanied by their plucky sidekicks, the friendly gut bacteria. Gorgonopsians took their place as the Olympians of carnivory.

All this happened on a planet where reptiles and other tetrapods were also proliferating. Some of the members of their groups had also evolved into giants, like the herbivorous pareiasaurus, but many still closely resembled their original 'lizard-like' body plan, the one they shared with the synapsids from the early days of terrestrialisation. This was true of one group in particular, the archosauromorphs, who would later take a leading role … but for the moment remained uninspiring extras on the Permian set.

Back in Scotland, we wound our way around Clashach Quarry that freezing November afternoon, searching the newly broken stone that had been blasted from the cliff faces. Trackways have already been set aside near the footpath for the public to see, ringed with gorse soldiers. These slabs, each containing one or more Permian footprints, were exciting, but had been found over 20 years ago when the quarry was at its most active. Did the rocks of Clashach hold more petrified Permian steps within their walls, or had those layers been removed?

In the 1990s a geologist named Carol Hopkins was researching the footprints near Elgin. Her work, which was sadly never published, was the last major investigation of the prints until I turned my attention to them in 2017. The historic finds and new prints found by Hopkins and subsequent professional scientists and amateur enthusiasts have been catalogued and kept safe in the collections at Elgin Museum, National Museums Scotland, and a handful of other institutions in Great Britain and abroad.

But Clashach has been neglected in recent years. The prints by the path are shadowed by foliage and caked in lichen. Most look more like geological accidents than biological traces. The interpretive sign by the amphitheatre that explained what kind of creatures once trod this shore looked as though it had been there since the Permian. On my second visit to the site, it had disappeared completely.

With her geological background, Hopkins had spent much time understanding what the rocks at Clashach could tell us about the environment of Permian Scotland. It was clearly an arid desert, composed of vast golden sand dunes. But the fact that there were prints meant some moisture must have been present to bind the sand grains together. Perhaps early morning dew, or even occasional rainfall – some rocks had blobs on them that could have been the impressions of raindrops on the parched landscape.

Hopkins observed that trackways were often made in sloping surfaces. This is revealed by the way the sand deformed beneath the animal's feet, leaving a larger rim on the downhill side of the print. Historical reports suggested that tracks were often travelling in the same direction, perhaps towards a water-source. At the end of the Permian, a body of water called the Zechstein Sea sat over what is now the North Sea and northern mainland Europe. Looking out from the sandstone quarry towards the Moray Firth, it seemed that geography, like natural selection, had undergone its own form of convergence.

The animals that walked this way, however, were not recapitulating, they were the first of their kind. Researchers moved on from tortoises, and believed instead that these footprints belonged to Permian therapsids. They suggested the harsh hot interior of Pangaea was a monoculture of the mammal lineage.

As more prints have been found around the world, the picture has been enriched. Examining similar sites in southern Scotland and Germany, researchers suggest there were at least five different types of animal responsible for the fossil footprints. They include therapsids, but also the other components of the Permian ecosystem, parareptiles like the large pareiasaurs and smaller-bodied kin. As well as this, burrows and insect traces – dimples made by multiple unidentified feet – are common. It is easy to picture a habitat not unlike the deserts of today, where animals coped with the intense heat of the day by pushing themselves under the surface or coming out after sundown. Maybe some of these morning-dew impressions mark not a trek towards water, but a return to shade after a long night foraging in darkness.

Clambering over the rubble at Clashach, our team hoped to find freshly exposed prints. If we were really lucky, perhaps we would discover something even more impressive. Since the Victorian period multiple quarries in the Elgin area had produced both footprints and bones from ancient animals from the Permian and subsequent Triassic. The majority of those quarries were long closed and seriously overgrown, eaten up by the landscape. At a site along the coast from Clashach, near-complete skeletons had been discovered in the 1880s, belonging to Permian dicynodonts and pareiasaurs. These have been referred to as the Elgin 'reptiles' – another moniker that needs a rethink, for as we now know, few of these creatures were actually part of the reptile family.

At Clashach, bones are exceptionally rare. Some broken fragments were all that had been found there until a chance discovery in 1997. In her study of the quarry, Hopkins had developed a good relationship with the quarry workers. She asked them to watch out for any new footprints that turned up during extraction, so that she could study them. But she also asked them to watch out for *holes*.

Sometimes, although rock is not suitable for preserving fossils, it can preserve their outlines. When animals died on the sand dunes of Permian Scotland they would have been buried, and the grains would have compacted around the body. But sand is porous, allowing moisture to pass through it. Over time this process can destroy bone completely, but if the surrounding sandstone is compact enough it remains in place and preserves the void where the bone used to be. Hopkins and fellow palaeontologists hoped that such fossil voids might be present at Clashach.

In the late 1990s, the quarry workers finally found what everyone had been hoping for. They broke open a boulder and inside it was a hole like a puckered wound, 10 centimetres (3.9in) across. Although it wasn't possible to see inside it, Hopkins inserted a piece of solder wire, which disappeared about 25 centimetres (10in) deep. It was a big hole and had some kind of complicated shape – it had to be a bone void. It was removed from the quarry, and using computed tomographic (CT) scanning and magnetic resonance imaging (MRI), the mould of a complete dicynodont skull was revealed inside.[*]

Our team was on the lookout for the next Clashach skull. Finding fossils can be serendipitous, but we knew that if it had taken 200 years

---

[*] A digital model was made of the void, and this was then made into a 3D model. You can see the void and the model in the displays at the Elgin Museum.

to find the first void at Clashach, our chances were vanishingly small. As expected, we didn't find a skull, but I did discover some new footprints. Little scampers across the sand no bigger than a coin, leaving only claw-marks behind. Local people and quarry workers are still finding prints there, and if we could discover something ourselves with only a few hours' searching, it was clear that there was still treasure to uncover from the site. This gateway to ancient Scotland had to be protected. Who knew when the next spectre of the Permian would raise its head from the sands of deep time?

Therapsids were the main innovators of the Permian. These many amazing groups made the world their home over 252 million years ago: emerging and flourishing, evolving new adaptations to hunt, digest, fight, climb and dig. Although they are all part of our bushy synapsid tree, so far all of the groups I have introduced you to are only cousins whose descendants, except in a few cases, didn't make it out of the Permian. We are related to them, but not as closely.

The group of therapsids from which our own lineage emerged were, compared to their contemporaries, almost unremarkable. They were close kin to the gorgonopsians and a similar group called the therocephalians, which included omnivorous and herbivorous forms.[*]

Our direct mammal line, however, was not much to write home about. Right at the very end of the Permian, they were late to the party. They were mostly quite small, and superficially there is little to tell them apart from the other therapsids. But to palaeontologists they are celebrities. This unassuming collection of animals gave rise to the ancestors of mammals. They are called cynodonts.

Cynodonts first appear in the fossil record of South Africa, but quickly emerge in other parts of Africa as well as Europe and Russia. They looked like compact dogs, miniature versions of their close kin the gorgonopsians and therocephalians. But by tracing the subtle tells in their bones, we know they were something new.

---

[*] There may even have been a venomous species, which would mark yet another first for the synapsid family, but this interpretation is not certain.

Their temporal fenestrae were larger, fitting increasingly enlarged and complex jaw muscles. Above the fenestra on the top of the head a crest had developed for those muscles to be inserted. This change was linked to chewing with more complex teeth, now fully differentiated into unmistakably mammal-like incisors at the front, canines and postcanines at the back. The dentary that held the bottom teeth dominated the jaw, while the bones at the back of the jaw became smaller still. Many of the changes that we saw in earlier groups went further among cynodonts, including modifications to the limbs, hips and shoulders. These foreshadowed later developments that set the mammal line apart from their contemporaries, creating a body plan that persists to the present day.

What do all these ancient skeletal changes really mean? For the palaeontologist, they are the Rosetta stone of evolution. But our bodies are so much more. The traits we most associate with modern mammals are of the flesh: hot blood, rich milk and ticklish fur. Therapsids may have been recognisably pre-mammalian in their underlying architecture, but what about the soft furnishings? What do we know about therapsid biology?

Although we say 'warm-blooded' or 'cold-blooded', these terms are actually over-simplistic. They refer to whether an animal maintains its body temperature internally or using its environment; but in fact, animals alive today have a spectrum of strategies for keeping themselves at the right temperature – and even the 'right' temperature differs between groups. The only warm-blooded, or endothermic, animals alive today are mammals and birds.* Pretty much everything else is cold-blooded, or ectothermic.

Ectotherms have little to no internally generated source of heat, so they rely on their surroundings to warm up or cool down. This strategy is ancestral for all vertebrates; in other words we all come from shared cold-blooded ancestors. All animals generate some warmth from everyday metabolism, but not enough to keep the body warm when temperatures fluctuate. The benefit of relying on the environment for heat exchange, rather than stoking your own fire, is

---

* Although some fish can show temperature control, such as the opah (*Lampris*).

that ectotherms can live frugally in terms of their metabolism: they need less food. Some crocodiles, for example, only eat a few times per year (but when they do, they usually gorge).

The down side to ectothermy is that you are limited by the temperature range of your environment. Some ectotherms can withstand larger fluctuations in bodily temperature, but only within a certain range. There is a good reason why there are no amphibians or lizards in the Antarctic, and very few insects. Only birds have adapted to remain on that frozen continent (and a few marine mammals, like seals). Another drawback for ectothermic tetrapods is that they can't maintain fast movements for long periods of time – they swiftly run out of steam.

Endothermic mammals and birds, however, can live almost anywhere, and are much more active. On top of their everyday metabolism, they generate extra heat from inside their cells. This is often done by burning fats and sugars, like coal in a furnace. This means they can remain active for longer periods, but the drawback is that endotherms are always hungry for more, needing to eat regularly to keep those fires burning. Every mammal and bird alive today is endothermic. This feature is tightly linked to their ecology: from how they feed to where they live, their reproduction, appearance and behaviour. It has been an entirely transformative innovation, achieved first in the mammal lineage, and later among dinosaurs.[*]

Figuring out when the first mammal ancestors became endothermic requires some detective work. Researchers look for clues in the bodies of mammals and birds today, the features they possess that are linked to being warm-blooded.

Mammals are furry and birds have feathers, this we learn as children. With a few exceptions like whales – and even they possess some stubble as babies – having an insulating body-covering is a feature shared by all endotherms alive today (fully aquatic mammals and birds use a layer of blubber to stay warm). Fur is rarely preserved in the fossil record because it isn't mineralised. Only in exceptional circumstances does a freak of geological preservation capture hair in deep time. The oldest unequivocal fur belongs to a Jurassic mammal called *Castorocauda*,

---

[*] Birds are dinosaurs. Despite insistence from some quarters that this is still a matter for debate, it is not. The evidence is overwhelming.

a character we'll meet in a later chapter. At only 160 million years old, this find pushes no boundaries – we are certain fur was present in mammals long before that time.

The evidence that at least some therapsids may have had fur is crap. Fossil crap in fact. In a quarry just east of Moscow in Russia, a team of palaeontologists discovered fossilised faeces, called coprolites, the size of cat scat. Inside they found a mixture of bone fragments, fish scales, bits of insects and various fungi and bacteria, suggesting it was deposited by a Late Permian carnivore with a flexible diet. Most excitingly though, some of the structures inside the fossil look like hairs.

Hair is composed of keratin, which is indigestible and so could theoretically survive a journey through an ancient gut. The structures in the Russian coprolite are a similar size and shape to fur, and could have belonged to whatever poor creature this little beast had recently eaten. This could constitute the earliest evidence that a few therapsids in the Permian already had at least some hair, even if it was just a tickly moustache. However, the structures might just be nothing more than random shapes in the minerals of the coprolite, or tricksy wisps of fungi or debris in the jumble of excreta.

The most obvious initial evidence for increased body temperature in extinct animals is a change in body architecture. As we've already seen, therapsid shoulders and hips had modified in such a way that their legs could be brought closer underneath the body, raising them further from the ground and making them more mobile. Increased activity, therefore, became physically possible. Their gait would have completely changed from their forebears', providing agility and manoeuvrability. These sorts of modifications go hand in hand with the emergence of a food chain built of hunters and the hunted. Being elevated from the ground also provided room to breathe, quite literally. All active animals need to breathe more to help fuel their activity: oxygen feeds their fires.

Have you ever looked up a dog's nose? Inside it is another clue to endothermy, one that is linked to their need for more air. When a warm-blooded animal breathes more it faces the same risk that the first tetrapods did back in the Carboniferous: water-loss. As we all know, we can survive for a while without food, but not without water, because dehydration stops your cells and organs from functioning properly. As warm moist air is breathed out during activity, heat and moisture is lost, posing a danger to the body. Mammals and birds have solved this problem through the development of ridges inside their noses called turbinates.

Nasal turbinates, or conchae, are like folded sheets within the nose. They are composed of bone or cartilage and covered with a moist tissue lining. There are two types of nasal turbinate: olfactory turbinates, which are used for scent-detection, and respiratory turbinates, which play a key role in breathing. As the outbreath passes over them they reabsorb the heat and moisture that would otherwise be lost from the body.

All warm-blooded animals on Earth today – except whales and a few diving birds – have turbinates, and research suggests that without them being endothermic is probably impossible. Supporting this, by comparing the total surface area of respiratory turbinates against metabolic rate, it is clear that the higher the metabolism, the greater the turbinate area, reinforcing the link between being active and the need to conserve moisture and heat.[7]

Applying these results to the fossil record, we see where turbinates emerge in the synapsid lineage. Olfactory turbinates for detecting smells are found in almost all synapsids, even the earliest pelycosaurs. Something akin to them is also found in almost all other tetrapods, which suggests that detecting scent in the air is among our oldest sensory adaptations, and emerged very early in tetrapod evolution. But respiratory turbinates are another story. It is not until the very last lineages of the Permian, the therocephalians and cynodonts, that we find possible evidence for their beginnings.

Although the turbinates themselves are too delicate to fossilise, the ridges that support them inside the nose can be seen. From close examination of fossil skulls using X-ray scanning, some researchers have argued that ridges on the inside of the nose in animals like *Glanosuchus* formed the base of respiratory turbinates. A carnivore that grew to two metres (over 6ft) in length, *Glanosuchus* is a therocephalian known from fossils in South Africa.

Not everyone agrees with this interpretation of the fossil, but if these ridges are the base of turbinates, then they comprise the earliest evidence for the emergence of this structure in the fossil record. It isn't long before they become a ubiquitous feature in every cynodont – a sure sign that the metabolism of our ancestors was changing long before the dinosaurs even reached the planning stage.

Another way to explore metabolism and endothermy in extinct animals is by looking at the internal structure of their bones. Skeletons are not dead – bone is a living, growing tissue that modifies and responds to pressures exerted upon it. Although bones vary a little among vertebrates, generally speaking they comprise a hard outer layer on the shaft, called cortical bone, with a less dense tissue called cancellous, or spongy bone, found at the ends. This is where most growth and modification takes place. Bones have a blood supply, which is greater in the spongy bone, and there is something in the centre called myeloid tissue – we refer to it as bone marrow.

Bone is made of a mixture of hard calcium phosphate and softer collagen. Inside bones are dedicated cells for growing new bone and breaking it down again. This happens naturally all the time, and also makes it possible to remodel bones to respond to stresses placed on them. One famous example is in the arms of professional tennis players: one study found that in whichever arm they favour, the cortical bone is around 40 per cent thicker than in their non-playing arm.[8] This is in response to the repetitive impact of hitting the ball, and also acts as reinforcement for their increased musculature. A similar study in 2017 examined the upper arm bones of female athletes and compared them to the same bones in women from archaeological sites. They discovered that between the Neolithic and the Bronze Ages, the intense manual labour of early farming lifestyles meant the average woman had upper body strength comparable to that of a modern athlete.[9]

When an animal dies, although the softer cells and vasculature (blood-vessel networks) of the bone decompose, the hard mineral parts remain. These microstructures capture some of the original structure of the bone, making it possible to study growth, healing and the stress the bone could withstand in life.

To examine bone microstructure, a practice called histology, the first step is to cut a sliver of bone around 0.03mm in thickness – thin enough to pass light through. This is the first hurdle for any histologist: trying to persuade museums and palaeontologists to agree to their precious fossils being chopped up. Some specimens are simply too fragile or rare for thin-sectioning. One alternative is to use powerful X-rays to see into the bone structure, but this can be costly and has physical limitations which we'll hear more about in later chapters.

Delving into the world of histology is like swimming in alphabet soup. There are so many terms for types of bone, locations within bones, bone cells and growth patterns – the result is both a mental and literal tongue-twister. Then there are the factors that affect what you see in bone-microstructure: the age of the animal, where in the body the bone is located, the environment of the animal, as well as modifications caused by fossilisation and how the thin-sample was prepared. Looking at pictures of histological thin-sections feels like a visit to a modern art gallery: it all looks interesting, but you're not always sure you get it (and if you think your five-year-old could have done it, you probably don't understand what you're looking at).

To figure out where warm-bloodedness emerged in the synapsids, researchers have sought evidence of growth rates in their bones. It was shown in the 1970s that there was a difference between the kind of bone found in earlier synapsids, and that of later therapsids. In the therapsids there seemed to be a lot more fibrolamellar bone (FLB), a fast-growing type that indicated they had a more rapid growth trajectory than their predecessors.

A study in 2007 led by Laëtitia Montes showed that there was a clear link between bone growth rates and metabolism,[10] supporting the inference that therapsids had faster metabolisms, and were therefore increasingly warm-blooded. Recently this finding was backed up by a closer interrogation of the bones of two anomodonts from the Late Permian, *Lystrosaurus* and *Oudenodon*. When compared with the earlier animals *Clepsydrops*, *Dimetrodon* and *Edaphosaurus*, the anomodonts showed much faster bone growth and therefore higher metabolisms – closer to the rates seen in some modern mammals with slower metabolic rates, like the grey mouse lemur, *Microcebus*.

A series of studies have also looked at the bone structure in gorgonopsians, and found similar evidence for fast growth. Many of these studies were carried out by Anusuya Chinsamy-Turan and her colleagues. Chinsamy-Turan, a South African researcher from the University of Cape Town, is one of the world experts in histology. Few people have sliced up as many ancient bones for science. Her book,

Forerunners of Mammals, brings together cutting-edge research on histology in the synapsid lineage carried out by her and her colleagues over the last few decades. It tells a complex, but fascinating story. Gorgonopsians, for example, seem to have shown signs of fluctuating growth patterns, with periods of faster growth, and times where growth ceased altogether. It's not a clear-cut picture, but suggests that this group of therapsids may have had a different metabolism from earlier synapsids.

Approaching the problem from a different perspective, one study sought evidence for warm-bloodedness not in bone structure, but in its composition. One component of bones and teeth is a phosphate mineral called apatite, and this contains oxygen, which is incorporated into the body from its surroundings. The nature of the oxygen isotopes changes with body temperature and things like climate and rainfall. Ectotherms, which control body temperature environmentally, have quite different patterns of oxygen isotopes compared to endotherms. This provides a method for examining fossils for clues to warm-bloodedness.

In 2017 a team of scientists obtained bones and teeth from therapsids as well as various modern reptiles and amphibians, and tested them for oxygen isotopes. Their results supported an increased metabolic rate in cynodonts; in other words, they were more warm-blooded. This doesn't mean they were like mammals today, but that their metabolic rate was higher than, for example, a reptile. They found less support for endothermy in other therapsid groups.[11] This suggested the first steps towards endothermy emerged somewhere between the end of the Permian and the Early Triassic.

Finding that fast-growing bone, FLB, in several groups of therapsids seems like an open-and-shut case for endothermy. But as ever in science, it's not so simple. Oxygen isotopes are conservative in the story they tell of therapsid warm-bloodedness, and not all therapsids seem to have had FLB, even though there are a handful of earlier synapsids that did. In the modern world this type of bone isn't entirely confined to endotherms either – it has also been found in some turtles and crocodiles. To confuse things further, it is absent in a couple of mammals and birds. It turns out that although FLB occurs in most animals that exhibit fast growth, that doesn't always mean they are warm-blooded.

This sketchy picture of endothermy might cast doubt that the first age of mammals was a hot-blooded one. But what is more likely is that endothermy itself, like so many of the features associated with mammals, emerged scattershot. It wasn't switched on like a lightbulb, lighting up all therapsids at once. Like the array of changes seen in therapsid body plans, increased metabolic changes probably occurred at different rates across the therapsid tree, from the most minimal increases in some groups, to larger bone-changing leaps in others. One thing we can say for certain, is that they were leaders in the world of radical physical change at the time. These creatures were undoubtedly evolutionary trend-setters.

That therapsids hit upon so many novel adaptations so early in the history of life on Earth is one of the major discoveries of the last 100 years. But as a group, they also pose a conceptual challenge. Evolution is messy. New ways of looking at fossils teach us that we can't trace a direct lineage back to our earliest ancestor, because we don't actually know who begat who in the genesis of biological life. Instead, we think of groups of animals like Russian dolls, each group nesting in a larger, more inclusive one, sharing common ancestors, until we reach the biggest of them all: our last universal common ancestor, or LUCA. Recent studies suggest it emerged almost as far back as the Earth itself, over four billion years ago[12].

We like our narratives simpler. We want one adaptation to lead to another, in an assembly-line through time. But the picture of major changes in evolution is less of an assembly-line, and more of a line-dance. Imagine that all the changes in the body of synapsids are like a row of performers holding hands. They are going for the world record in endothermy, and to get there the whole line has to reach the other side of the dance hall. A particularly enthusiastic dancer runs for it – but is held back by the hands of the slower movers that flank them. Only by waltzing together can the entire dance troupe reach the far wall and take the prize.

This is the analogy given by one of the most prolific authors and researchers of therapsids, Tom Kemp. He called it correlated progression, and used it to visualise something we often struggle to accept: that there may not be a single clear accumulation of traits leading to warm-bloodedness – or any other feature – but a list of separate changes that all occur at different rates in different lineages. As the changes accrue, they lead to a wholesale different organism, the other side of the dance hall.

It's a good analogy, but it might suggest that all the dancers *want* to reach the other side of the dance hall. Natural selection has no goal, but rewards and culls like a tyrant, based on the whims of survival. It might be better to imagine that line of evolutionary hand-holders was comprised of five-year-olds, playing musical chairs in the dark.

✇✇✇

Quarries continue to provide some of the best opportunities to see into deep time. Not only did many of our first fossil discoveries come as a consequence of digging for building materials, but they continue to do so. Around the world, new fossils are still turning up as we create gravel, pull out paving and (unfortunately) continue to extract coal.

For those who appreciate unspoilt nature, quarries are not places of beauty. They are scars on the landscape, seemingly at odds with the natural world. It can be hard to reconcile this as you watch the mechanical arm of a digger thrusting into the rock face, destroying it like a child stomping on a sandcastle. But these sites of destruction are also the sources of construction – not just of buildings, but of knowledge.

To find fossils we need exposures of fossil-bearing rock, and such faces are surprisingly sparse. Where the right sedimentary stones do come to the surface they are mostly coated in a lush pelt of foliage. The Earth herself is like a giant green mammal, bristling with life. It is often only when our industry drives us to dig into her flesh that we find the secret bones of lost worlds are in there, under the skin. Without digging, we would know almost nothing about her – or our – history.

As we strode back up the hill to our cars, I looked behind into the empty bowl of Clashach Quarry. We had left our footsteps everywhere, winding trails through the damp sandy floor. What would a future ichnologist make of these jumbled circuits? We'll never know. All we can say for certain is that sites like this and many others around the world must be protected and studied if we're to keep exploring our distant past and the lessons it can teach us about ourselves, and the beasts that came before us.

CHAPTER SIX

A Total Disaster

As a paleontologist, I'm familiar with extinction.

Farish Jenkins, after being diagnosed with cancer[1]

Dogs barked. Cars and crickets hummed and grated. Above me tiny bird-feet clitter-scratched across the tin roof of my B&B. Next door, another guest was watching TV soaps with their door wide open. Afrikaans drama leaked out onto the gentle breeze that rose as the sun went down. *Go away! Go away!* cried the go-away bird – or is it because the whole city wishes they would? My nose wrinkled: woodsmoke and burnt coffee. A tar-black lizard stopped to watch me, this pale skinny mammal sitting on the veranda. *The sun is setting, why are you still trying to bask?* It shook its head and ducked under the ivy.

All of the grass was yellow when I was there, but I'm told Johannesburg is lush and green in summer. I was glad to arrive in the early spring, without the heat or insects to greet me. When I stepped off the plane that morning after an 11-hour overnight flight from London, it was going to be hotter back in Britain than it was in the South African city. I was relieved to escape another oppressive English late summer day.

The next morning I would join a team from the University of Witwatersrand to do some fieldwork. Our crew of international collaborators would be taking a seven-hour road trip south to work with the local people in a small village on the border with Lesotho. There, the fossils of dinosaurs, giant crocodiles and mammal predecessors have been found emerging like zombies from the dry veldt.

The Triassic was calling.

This time period, which followed the Permian, witnessed a radical re-write of the world's fauna. The beasts that flourished in the first age of mammals were replaced almost wholesale by reptiles.

Despite over 100 million years of diversification and ecological innovation, therapsids were swept away in a mass extinction event so brutal that life on Earth came closer to annihilation than ever before or since.

Thanks to the demise of the non-bird dinosaurs, it is the end-Cretaceous that is the Hollywood superstar of mass extinctions. It happened around 66 million years ago, but everyone is still talking about it. Killer asteroids are certainly headline-worthy, but if death was money it wouldn't be the highest earner. If you want the real Dwayne Johnson* of mass extinctions, you need to check out the end of the Permian.

The end of the Cretaceous was murder from space, but the end-Permian was filicide.

An epic 252 million years ago, the first great chapter of evolution on Earth was brought to an apocalyptic halt. At least three-quarters of life died.[3] The next geological period, the Triassic, began with a world virtually wiped clean. Things got weird, *really* weird – I'm talking sprinting bipedal crocodiles and platypus-faced aquatic reptiles. There was a complete changeover of animal groups, and the modern ecosystem was birthed from the rubble, including mammals as we know them.

It was the Earth herself who eradicated her inhabitants at the end of the Permian. The evidence of her crime is found in the landscape and chemistry of the rocks of Siberia.

The broad sweep of the Siberian heartland stretches eastward from the flanks of the Ural Mountains. This landscape forms a series of step-like hills called the Siberian traps.

The rocks are volcanic, but these are not from the kind of volcanoes most people imagine. We think of cone-shaped mountains blowing their tops off in explosive moments and raining debris at their feet. But the volcanoes of ancient Siberia were *flood* basalts. Rather than

---

* Johnson topped the Forbes annual list of the highest-paid actors in 2019, earning $124 million in one year. The fact that he is known as 'the Rock' makes him an especially delightful example for this metaphor.[2]

sudden upward bursts, these eruptions are like pots bubbling over on the stove. Runny magma from beneath the Earth's crust pushes up through cracks. These are often caused by plate-tectonic shifts over the top of active plumes of molten rock in the Earth's mantle. They are less explosive, but the lava is more liquid, and so pours out across the landscape and smothers it for hundreds of miles.

The scale of the Siberian trap flood basalts is almost incomprehensible. Over the course of a few million years and multiple phases of eruption, as much as seven million square kilometres[4] of the continent disappeared beneath the volcanic spew. The volume of lava is hard to determine, but scientists estimate well over three million cubic kilometres of the stuff ended up on the surface. To put that in perspective, if you were to pile all of this flood basalt on top of China, it would cover the whole country in a layer 300 metres deep.

This wasn't just a bad time to be Siberian. Of course the eruptions would have killed almost all life in the immediate area and produced enough ash to have an instant effect on the climate. But volcanoes produce much deadlier products with global reach. Sulphur-rich gases, methane and carbon dioxide were released in vast quantities, and these severely affected the climate.

Living in a world currently heating up from man-made climate change, we're all too aware of the impact of carbon dioxide (CO_2) on global climate. It traps heat, stopping it from escaping into space, and so warming the atmosphere – the aptly titled 'greenhouse effect'. We are currently in trouble thanks to a rise in CO_2 levels to over 414 parts per million (a precipitous increase from around 280 parts per million for the last 10,000 years). Imagine how much worse it was in the end-Permian, where concentrations sky-rocketed to over 2,000 parts per million (estimates vary, with some suggesting nearly 2,500 parts per million).[5] Although CO_2 levels naturally fluctuate through time, this is higher than the very highest concentrations seen at any other time in Earth's history. For example, in the Devonian it reached as much as 1,500 parts per million, but this was still way below the end-Permian. Also, these changes usually happen relatively slowly, providing time for the living world to adjust. But like anthropogenic climate change, the end-Permian heating was too rapid for natural selection to keep pace.

The steroidal rise of CO_2 was enough to drastically impact life on its own, but the second horseman of this apocalypse, sulphur, really stuck the boot in. As they linger in the atmosphere, sulphate particles attract water vapour, and together they condense into clouds. When the sun hits them, it dissolves the sulphates to form sulphuric acid, which then rains on the world like a brutal facial peel. The plants already struggling to photosynthesise under the thick blanket of ash in the heat of global warming are then seared by the very rains that should wash and sustain them. Entering waterways and seeping into the sea, acid rain dissolved the shells of molluscs and plankton, undermining the entire food chain.

The end of the Permian was the biggest mass extinction of all time, and is sometimes called the 'Great Dying'. Around 81 per cent of life in the sea was wiped out, and 75 per cent of life on land (figures vary, but these are solid estimates for losses at species level).[6] The list of groups that didn't make it through is immense, including entire families of corals, all of the iconic trilobites, giant sea-scorpions and multiple groups of fish that had persisted for over 150 million years since the Devonian. On land, every group was hit hard: amphibians, reptiles, and therapsids especially, felt the blow. Even insects struggled, with as many as nine entire orders sinking into oblivion – perhaps the only mass extinction of insects known. Most of the therapsid groups perished entirely, including those perfectors of the herbivore-carnivore food chain.

There are multifold lines of evidence supporting our knowledge of the Permian mass extinction. The Siberian traps comprise the first, both by their presence and their date. Volcanic rocks contain a mineral called zircon, which can be used in radiometric dating thanks to the uranium it contains, which decays at a known rate. The zircon dates tell us that eruption of basalts began around 300,000 years before the end of the Permian, and continued for at least 500,000 years into the Triassic. This smoking gun is accompanied by the signature of geochemistry in rocks from around the world. In places like Meishan in China and Siusi in northern Italy, geologists have found huge changes in carbon and sulphur isotopes across the Permian–Triassic boundary, showing that the natural cycles of these elements were radically disrupted.

The composition of chemicals in rocks and fossils tells us that the oceans were not only more acidic, but became anoxic – depleted of oxygen. This creates dead zones where little can survive. Ocean anoxia happens when the climate warms, and it results in disruption of food chains and ocean circulation – we see these events in the modern world due to anthropogenic climate change. At the end of the Permian, huge swathes of the once-rich Palaeo-Tethys were turned to a hot deadly soup, suffocating marine life. The temperature of the water soared to up to 40° Celsius (104 Farenheit) at the equator – that's as hot as your evening bubble-bath.

Fossils also tell us how the impact was felt among animal groups. We see clearly that widespread and successful lineages disappear from the rock record almost instantaneously in geological terms. Where life tried to hold on, things were tough. The forests of higher latitudes, such as the expansive coniferous woodland that covered parts of what is now Europe, disappeared. In the remains of organic sediments there appears to be a spike in the numbers of fungal spores, presumably being released as vegetation decomposed en masse. Rot ate away at the Earth.

The destruction of the end-Permian extinction event was on a scale beyond comprehension. The impact was protracted, occurring in cumulative bouts over a million or so years. But the consequences were even more far-reaching. It would take 10 million years for life to reassemble into something approaching the complexity of the therapsid-rich ecosystems that preceded it.

The world that emerged from the volcanic dust was unlike anything that came before. Understanding how life survived and what happened over the next 50 million years of the Triassic is a mystery palaeontologists are still solving. Most curious of all is why the mammal lineage, so successful and diverse in the Permian, relinquished their prime spot to the reptiles. The repercussions of this change, however, would be the assembly of mammals as we know them today.

Our truck pulled up off the road, sending up a billowing cloud of dust. Four of us stepped onto the dirt of the verge, squinting in the sunshine. Ahead, the road continued across a low bridge, and the veldt stretched beyond to the foothills of Lesotho's Drakensberg. Behind us the village of Qhemegha was hidden by a low rise. A dense network of water-cut troughs and valleys, called dongas, networked across the plain to our right. It looked like an image sent back from NASA's Mars Rover. The parched and reddened landscape was sore from winter drought.

We walked down the side of the bridge onto a flat riverbed as wide as a swimming pool, but as dry as a biscuit. The intense rainfall that cuts up this landscape was absent, but its imprints remained like a trace fossil. The rock was worn smooth and the sides of the riverbed were collapsing, the characteristic erosion caused by raging water.

'Is there any chance of flash flooding?' I asked my fieldwork companions. Coming from a country perpetually moist as a wet-wipe, all I knew about dry river beds was from movies and nature documentaries. I trusted the two South Africans and a Swazilander to know the landscape and its dangers. They exchanged glances. 'Probably not,' they replied, and walked off downriver. I eyed the base of the bridge, which had large chunks of stone and building debris wedged between its struts, hammered in like square pegs in round holes. *Probably not.*

The empty riverbed widened, then disappeared off a shelf. The layer of rock we were standing on ended as abruptly as a table's edge. We walked together to the lip and peered down. Car-sized slabs of stone lay jumbled around a teal pool of water. From a hidden spring in the rock face water trickled in, and the whole riverbed meandered away at the bottom of a steep-sided ravine at least four storeys deep. In the rainy season we would be swept over the rim in a spectacular waterfall. The action of the water was steadily cutting backwards, causing this waterfall to moonwalk its way across the veldt.

'That's where we're going,' announced Lara Sciscio, pointing into a gaping hole in front of us. She tightened the string of her hat under her pointed chin. Her colleague, Michiel de Kock, scratched his stubble and assessed the descent. He nodded, 'I'll get the drill.' Sciscio and de Kock had a crucial job here during our fieldwork. For the rest of us our top priorities were finding new fossils and securing the previous years' fossil finds for transport to the laboratory. For Sciscio and de Kock it was about the rocks themselves.

The rocks around Qhemegha are part of a massive geological sequence called the Karoo Supergroup. This is found across most of South Africa and Lesotho, and surfaces in parts of the surrounding countries of Malawi, Namibia, Swaziland, Zambia and Zimbabwe. As well as covering an area bigger than Germany, these deposits are many kilometres thick and span 150 million years of geological time.

The oldest outer fringes of the Karoo Supergroup are Carboniferous in age. Then there is a vast depth of Permian rock splayed out across the body of South Africa. Finally the Triassic rings a bullseye of Jurassic lavas, perched on top of it all like a cherry on a layer cake, and forming the country of Lesotho. Most of the younger rocks that formed on top of this in the intervening millennia have been scoured back and washed away.

We were walking where the Elliot Formation met the underlying Molteno Formation. Both layers of rock have long been known to be Late Triassic to Early Jurassic in age, based on traditional methods like faunal succession: looking at the species of fossil animals and plants preserved in them. They capture a crucial moment in evolutionary history.

Life had found its feet again after the total destruction of the end-Permian, but then there was another, much smaller mass extinction at the end of the Triassic.* The scale wasn't so apocalyptic, but still enough to re-structure the ecological landscape yet again. This key moment in animal evolution remains poorly understood – we don't know how or why some groups vanished while others made it through. To figure this out, we need more fossils.

The ages of the Molteno and Elliot formations are currently not precisely known. These rock groups cover a huge area, dipping in and out of valleys and peeping from beneath Jurassic hillsides. It was likely that the actual dates of the layers varied on a local scale, meaning that the different fossils found around Qhemegha could be as much as 20 million years apart in age. It would be difficult for palaeontologists to understand what their finds told them about the changeover of animal groups in this part of Africa if they had such poor chronology for their discoveries.

---

* To be precise, there were perhaps two mass extinctions at the end of the Triassic. A slightly smaller but significant event occurred in the Carnian, called the Carnian Pluvial Episode.

Sciscio was there to help solve that problem. She and de Kock were using a drill to take rock samples from points throughout the Molteno and Elliot formations to determine their age using palaeomagnetism (detectable changes in the Earth's magnetic field) and radiometric dating.

For sedimentary rocks, most radiometric dating techniques are unreliable because the stone is composed of grains from older rocks that have been eroded and recombined, as we saw in a previous chapter. But there is some useful information we can gather. Some sedimentary rocks contain grains of zircon – the same mineral that allowed precision-dating of the flood basalts in Siberia. The zircon contains the radioactive elements uranium and thorium, and their decay can be measured to give an estimated age for when the original volcanic rock erupted and cooled. It was only after the eruption that erosion and secondary deposition (the second time the rocks were deposited somewhere) took place. Logically, these radiometric dates give us a maximum age that the sedimentary rock could have formed – after all, rock can't be deposited before it erupted.

Sciscio had already worked out the relative ages of a section of the Elliot Formation at another location.[7] She would match the results from Qhemegha with her previous results to work out how the two sites were related to one another. Eventually, this work will give scientists a picture of the deposition of rock across the entire region, providing a framework upon which palaeontologists can hang their discoveries.

I watched Sciscio's small frame as she clambered fearlessly over the lip of the empty waterfall, disappearing into the steep ravine below. My task was up here. Along with fellow palaeontologist Kathleen Dollman, I set off for the crumbling river banks. While our friends drilled rock, we would comb the eroding sediments in search of fossils.

↓↓↓

In this part of South Africa the ground is full of bones, and the whittling of the rainy season exposes them piecemeal. Searching is thirsty work. It was winter and although cold by local standards, it felt warm to my blood, hovering between 15 and 20° Celsius in the daytime. The air was so dry that when I brushed my hair the strands cracked and broke in handfuls, and static made it look like I'd stuck my finger in a power socket.

The first day searching is exciting. You pick up bone after bone from the purple-grey weathering surfaces and think yourself successful. But single bones are common, often badly damaged and rarely worth collecting. What scientists really hope for are complete jaws, skulls and skeletons, things that can be analysed to tell us about species diversity and how animals moved and lived. These specimens are less easy to locate, but we knew from the local people we worked with – who continue to find new fossils and localities to search – and from previous years spent in Qhemegha, that they were there. Dollman headed straight for the river bank beside the dry waterfall. Rather than cover the same patch, I decided to walk on to the far bank and search the dongas cutting into the hillside.

The Elliot Formation, previously known as the 'red beds', is mostly made up of mudstone and siltstone. The rock is soft and gives easily with a gentle tap from the hammer, raining down in a bruised spectrum of purples and reds. This area was once squiggled with meandering rivers, bringing life to the plain and steadily dropping their burdens of mud. Sometimes they also carried bodies, laying them to rest in a quiet ox-bow for us to exhume 200 million years later.

After an hour I stood high above the river, looking into the ravine as I camel-gulped water. Below me, Dollman was waving her arms. I put the bottle away and strode back to join her. While I'd been walking up and down the crumbling cuts fruitlessly, she had barely moved more than a few metres. As I arrived I discovered why: bones were poking out of every mound. I traced one back into the loose earth with my geological hammer and it disappeared deep into the hillside. It was an assemblage of them, and they looked in good condition. She'd struck palaeo-gold.

Pulling out her phone, Dollman sent a message to our group. The rest of the team were at another site jacketing and removing the bones found last year. It wasn't long before a second truck parked beside ours and a familiar figure strode down the river bed. It was the leader of the project, Jonah Choiniere from the University of the Witwatersrand in Johannesburg. Originally from the US, Choiniere is one of the world's leading palaeontologists and a steadfast supporter of the next generation of South African researchers. In his dirt-caked jeans and faded T-shirt, he looked less like a scientist than a farm-hand, but he had the manners of a polite cowboy. The sun was already turning his features toast brown – or perhaps it was just 10 days of showerless tent-life? One of the other palaeontologists on the team joked to me a few nights previously that Choiniere enjoyed fieldwork so much because it allowed him to indulge in his two greatest passions: smoking and not washing.

Choiniere arrived with a rolled-up cigarette and his trademark air of calm. 'What have you got?' Dollman showed him the bones, and together we evaluated the scene. He picked up a few of the complete ones we'd carefully removed and placed on a nearby rock. He pinched his lips and nodded appreciatively. I couldn't see his eyes because they were hidden by aviator glasses, but he was clearly pleased. 'This looks like a sauropod, but this one' – he picked up a limb bone in three fragments, and held them together in anatomical position – 'this looks like rauisuchian'.

'Is that good?' I asked, being less familiar with reptiles. Choiniere gave an understated nod. 'Oh yeah, that's good.'

The fossils at Qhemegha were first found by a local shepherd, Dumangwe Thyobeka. He frequently visited us as we excavated to chat and find out where his discovery had led. He was tall, and usually wore a baseball cap over his short dreads. When I shook his warm hand it was dry and hard, like the stem of a shrub.

The area is traversed by flocks of sheep and cattle, and herdsmen like Thyobeka walk for miles daily as the animals forage. In winter it's a dusty, hungry business, and it isn't just the cattle's impressive horns that protrude; their ribs also jut out like xylophones. Sometimes we would find a cow skull or limb bone in the dongas, evidence of hard times for the village.

Thyobeka found the first bones when walking to the graves of his great-grandparents nearby. He knew his discoveries were too big to belong to cattle, and showed them to fellow villagers Sginyane Ralane and Themba Jika-Jika. Ralane, also a village elder, knew exactly what he'd found: fossil bones. He'd read about dinosaurs and flying reptiles, and always hoped to find some. Together they realised the bones belonged to dinosaurs, and that their presence would undoubtedly interest scientists, and could be a significant boost for the community.

The village elders sought advice, and ended up in contact with Choiniere at the Evolutionary Studies Institute at Witwatersrand. Now Choiniere and his team worked with the villagers and colleagues from universities across South Africa as well as Europe and America. Together they were excavating and studying the fossils on behalf of the community of Qhemegha. The researchers openly shared all new discoveries with the village, welcoming daily visits to the dig site and learning some of the vast local knowledge of the landscape. At the week's end there was a slide show where Choiniere's team talked about the science of palaeontology, and answered questions about the fossils, evolution and geology. Ralane joined our dig most days, excavating and passing along his expertise. He had a keen eye, having made many significant discoveries over the last few years. He showed some of them to us in his garden, where delicate pink plum blossoms bloomed above rows of vegetables.

Local councillors and dignitaries visited the village too, and there were discussions about how the work taking place there might provide economic benefits for the area. It was still early days in the research, but the site looked as if it might be one of the largest bone beds in South Africa, and a real asset for the community.

At the start of the week when our team arrived at the main excavation site, I was greeted by the sort of scene most people imagine when they picture a palaeontological dig. The dry landscape, a team wearing checked shirts and T-shirts, wide-brim hats and sunglasses. The stereotype was set in that iconic early scene in *Jurassic Park*, where the palaeontologists are uncovering a complete *Velociraptor* skeleton in the North American Badlands. There are some truisms: landscapes like this erode readily to expose bone, and clothing is chosen for practicality against sun and hard wear. But our teams were not all rugged outdoorsmen, and there would be no dainty brushes delicately swishing sand aside to reveal perfectly articulated skeletons. That's just Hollywood stuff.

I dropped the bag of hammers and chisels I was carrying with a loud clink. In front of me at the dig site were two enormous plaster casts that looked like giant dirty marshmallows, beige with grime. Two metres across, they sat in human-made pits, thick-bottomed with sediment from the rainy season. The tattered edges of burlap peeked through, flapping in the hot wind.

The previous year, the team had honed in on the skeleton of a dinosaur, and a cluster of smaller bones likely to belong to one of the few therapsid lineages that made it through the mass extinction, the dicynodonts. They had excavated around them, leaving the fossils encased in their graves. As time ran out they covered them in a protective layer of rough sacking soaked in plaster of Paris, and hoped these hardened shells would protect them from the elements until they returned.

We would be spending the first week of fieldwork widening and digging out the pits, before washing the old plaster and applying new layers. In the second week, someone with a digger was coming to flip the multi-ton marshmallows over so we could plaster their undersides. Then they'd be lifted onto a flatbed truck for the seven-hour journey back to Johannesburg. In the laboratory at the University of the Witwatersrand, a team of South African fossil preparators were ready to work on the jumbled contents, finally freeing them for study.

The bonebeds of Qhemegha could prove to be the best locality in South Africa for Triassic fossils. All we needed now was to collect them.

The Triassic is the overlooked sibling of the three periods of the Mesozoic era. She lacks the glamour of the Jurassic or the self-assurance of the Cretaceous. She is an introvert elder sibling that no one really understands. But if you take the time, she definitely has the best conversation of the three.

The Triassic is full of new ideas. This time period, spanning 51 million years, begins in devastation at the close of the Permian, and comes full circle to end in another mass extinction. The intervening years are eventful, and their study has given us unprecedented knowledge about how life recovers from the brink of annihilation.

We'd like to think mass extinction is rare. The Victorians struggled to stomach the reality of it, in a world forged and cradled by God. Now, though, we are acclimatised to the disappearance of species over millions of years. Some of us are appropriately shame-faced about recent losses due to human activity, but humans have an uneasy complacency about such oblivion. When death reaches a certain scale, humankind goes a little numb. The comedian Eddie Izzard once joked that when deaths enter the hundreds of thousands, we are so dumb-founded that our reaction can be nonsensically blasé.[*]

Palaeontologists, who study evolution's corpses, are both desensitised and exquisitely aware of life's fragility. We are just living between extinctions, in a pause between planetary heartbeats.

At least 20 mass extinctions are known from the fossil record. Which ones are considered 'major' is a matter of statistics. In 1982 two palaeontologists named Jack Sepkoski and David Raup looked at the numbers of groups of marine invertebrate animals in the fossil record. They compared how many appeared and disappeared through time against the normal background rate of extinctions. On average, species only tend to last between one and four million years. The rate they appear (called speciation or origination) and are snuffed out again rises and falls like steady breathing.

Sepkoski and Raup could see from their data that there were five points in the fossil record where the number of extinctions rose significantly above the number of originations – a sharp gasp in the pattern. The end-Permian event was by far the largest, followed by the end-Triassic and end-Cretaceous events, and then two further back, one in the Devonian and one in the Silurian. These are the classic 'big five'. Recently more perishing moments have been added to the list. There was a nasty blip 10 million years before the end of the Permian, and another identified in the Triassic. And of course there is one happening right now, caused by a particularly short-sighted ape.[†]

We are often preoccupied with identifying the drivers of mass extinction. These can include massive volcanism, asteroids, sea-level change, climate change and plate tectonics, usually a combination of several of these, altering the environment too quickly or radically for established life to cope. Although we all naturally want to know the

---

[*] In one of Izzard's classic dark skits, he tackles our difficulty comprehending genocide with trademark whimsy, joking that for someone to accomplish the murder of thousands they must have to get up quite early in the morning.

[†] Of the many excellent books dedicated to our current crisis, I must recommend *The Sixth Extinction: an Unnatural History*, by Elizabeth Kolbert.

cause of the malady, what can be more tantalising is understanding the recovery.

One realisation resulting from Sepkoski and Raup's work is that despite the rhythmic destruction, we are still here to tell the tale. The story of how life gets back on track is not merely a curious evolutionary story; it helps us appreciate the full impact of our actions as the precipitators of the latest mass extinction, and to anticipate the response of the natural world in coming centuries.

In the aftermath of end-Permian mass extinction, the few tetrapod groups that survived didn't have it easy. A massive dead zone stretched around the equator, where land and sea temperatures made it almost impossible for anything to set up home. There is a 'coal gap' in the rock record at this time because there were no forests to create the deposits to form it. Similarly there is almost no coral, or the rock called chert which normally forms from the steady downpour of silica-rich plankton on the sea floor.

The extreme conditions and loss of complex habitats across the globe had strange effects on diversity. Those groups that made it through, striding out of the flames like unkillable Terminators, did well in the aftermath – but not necessarily in the long run. These short-term thrivers are called *disaster taxa* (taxa is the plural of taxon, a group of organisms). The classic example in all the textbooks also happens to be a member of the mighty synapsid lineage. It is called *Lystrosaurus* and belongs to the dicynodonts, whose beak-and-tusk combo set the group apart from other therapsids in the Permian.

If you go digging in the bone-beds of the earliest Triassic, you can't help but find *Lystrosaurus*. These determined herbivores were everywhere, thriving from Pangaea's Laurasian head to her Antarctic feet. They are so widespread on the now-separated landmasses, that their presence in the fossil record across multiple continents provided key evidence for the understanding of plate tectonics and the existence of the supercontinent of Pangaea.

Lystrosaurus species ranged from cat-sized to cow-sized. All of them were rotund and short-legged, retaining a semi-sprawl, unlike the more upright stance of other therapsids. Their tails were short – a stubby triangle on a waddling behind. Snuffling through the undergrowth, they took advantage of the herbaceous vegetation that had sprung up with the disappearance of Permian forests. Softer, low-growing plants without woody stems, like quillworts, spike and club mosses and lycophytes, fell to the secateur-beak of these dicynodonts. Up to 90 per cent of all vertebrates alive in the earliest days of the Triassic were *Lystrosaurus* species. It is one of the only times we know in history that one small group of tetrapods so overran the Earth – a veritable plague.*

The organisms that make up post-disaster ecosystems tend to be generalists, eating a wide range of foods and adapting to varied environments. It is likely that *Lystrosaurus* wasn't a picky eater. The remains of its burrows suggest that it could have taken shelter from the harsh heat and acid rains, further assisting survival against the odds.

It did not live alone on this recovering planet. *Lystrosaurus* was accompanied by a straggly band of fellow therapsids, the therocephalians and cynodonts. These survivors were decimated in number, but nonetheless clung on. Some ancient amphibian relatives made it through as well, recovering rapidly to become aquatic predators. A scattered remnant of the reptile lineage also staked their claim to land: relatives of the giant Permian pareiasaurs and the ancestors of all living reptiles and archosauromorphs shared the cleansed Earth. At first they were a relatively small component of the ecosystem.

The dicynodont monoculture was not to last. The key feature of disaster taxa is that they are short-lived. For a time the end–Permian extinction event reduced ecological diversity and many ways of life were not being exploited. Even after 15 million years of rehabilitation there was a lack of large-bodied herbivores or carnivores, niches occupied in the Permian by therapsids like gorgonopsians, and sauropsids such as pareiasaurs. Along with a lack of the very large, there was a paucity of the very small: fish- and insect-eaters smaller than a cat took time to re-emerge. But whereas this diversity of lifestyle and body size had taken over 100 million years to evolve the first time around, it happened a lot faster in the Triassic because the generalist survivors of the mass extinction provided a leg-up. There was no need to start from scratch; natural selection already had the foundation to build the next complex ecosystem.

---

* Humans constitute a similar imbalance. So do chickens.

Over the first 20 million years of the Triassic, ecosystems re-established themselves. The mutations that flourished in the Triassic created an unprecedented new world. This recovery has taught scientists that when multiple animal groups disappear, it creates myriad opportunities. This relaxes competition for survival between groups, giving them space to explore new ways to make a living.

A similar evolutionary mechanism worked its magic on a smaller scale in the Galapagos, as Charles Darwin began to grasp after his iconic voyage on the ship HMS *Beagle*. In time, he saw that the finches that had reached the Galapagos were the progenitors of all the diversity of finches thriving there today. Finding themselves on the recently erupted islands, mutations that provided a novel edge in the fight to compete against one another stacked up, widening some bird bills for crushing seeds, lengthening others to poke holes in cacti and nibble their insides.

The therapsids had made it through, but after the explosion of *Lystrosaurus* wound down, it was the reptile lineage who were first off the mark to take advantage of the opportunities around them. Like teenagers leaving home for the first time, they started to experiment. This resulted, among others, in some of the most hyped and beloved extinct animals in the fossil record: the dinosaurs.

Dinosaurs have come to characterise the Mesozoic, but they were actually slow starters. As the Triassic got going, different reptiles began exploring land, sea and sky. In the water, plesiosaurs and ichthyosaurs were flattening their paddles and making their home in the depths. These marine reptiles are sometimes mistaken for dinosaurs, but their lineage was long separated, at least as far back as the middle of the Permian.

One of the strangest Triassic marine reptile groups are the hupehsuchians. Their fossils are found in China from deposits laid down in the edges of the Palaeo-Tethys Ocean. They are almost crocodilian at first glance, but bestowed with oversized paddles for limbs. Running down the spine from shoulder to hip are bony nodules arranged like speed bumps. At one end the heavy tail tapers to a dagger-tip, while at the other sits a preposterously tiny head. Hupehsuchians were toothless, feeding on soft invertebrates like marine worms. They ranged from no bigger than an otter, to porpoise-sized. One of them evolved a skull that to all intents and purposes was a trial run for the platypus, complete with a flattened sensitive 'beak' and tiny eyes.

Elsewhere other reptiles were branching out. The ancestors of turtles lurked nearby – at first on land, not getting their feet wet until later in the Triassic. Among the archosauromorphs two leading groups emerged: the crocodile line and the dinosaur line. Both began the journey as superficially lizard-like creatures, but the Triassic gave them space to play games with their body plans. Near the base of the dinosaur line lie their close relatives: the first vertebrates to evolve powered flight, the pterosaurs.

In the Triassic, these diverse reptile groups beat therapsids in the race to reclaim niche-space. The earliest ancestors of dinosaurs were mostly riffing on a theme at first: to the non-specialist they all look like upright lizards with long necks. A few of them figured out how to walk on two legs, a specialism capitalised on by their descendants (and proving that apes are nothing special). They were the precursors of theropod and ornithischian dinosaurs, some of whom returned to all fours: animals like *Megalosaurus* and *Stegosaurus*. Another group called the sauropods would become super-sized. They are famously represented by the long-necked mega-tonne herbivores like *Diplodocus* and *Brachiosaurus*.

But all of this was still to come. The dinosaurs would go on to survive the end-Triassic mass extinction, and become one of the most important animals in the Mesozoic world. In the South African Karoo, their precursors provide a key to the age of different rock beds.

At first, however, it was not the dinosaurs who took the lead on land in the Triassic.

We think of crocodiles today as lumbering danger-logs: strewn on river banks for months without eating, then drowning a hapless thirsty animal in a short-lived frenzy. But in the Triassic, croc-relatives were among the most prolific and interesting animals around. They included small greyhound-like runners; these were slender and long-limbed, and some may even have been bipedal. Other crocodile-line relatives were huge, such as the 4.5-metre-long (15ft) horned herbivore *Desmatosuchus*, which looked like the love-child of a komodo dragon and longhorn cattle. Later in the Mesozoic the crocs continued to diversify, evolving fully marine super-predators and even herbivores, styled after the fashion of the piggy *Lystrosaurus*.

One of the early groups in this prolific croc lineage were the rauisuchians. They are well known from Europe, North America, Russia, China and Argentina, but less so from South Africa. Rauisuchians are the poster-children for the Triassic – indeed their presence in any strata is indicative of the age of the rocks. They had taken over from the gorgonopsians as the top predators of their time. Like gorgonopsians, this reptile group had a more upright stance that provided speed and agility in hunting. Looking at their skeletons, it is as though someone stuck the head of a *Tyrannosaurus* on the body of a tiger. They undoubtedly fed on the ancestors of dinosaurs in a reptilian recreation of the Permian therapsid-led food chain. The jobs remained the same, even if the employees were different.

Although rauisuchian bones are strewn in the mudstones of the South African Karoo, most of their fossils are broken and poorly preserved. Many of the fragments recovered in the past have little to no information about where they were found, limiting their usefulness for research. But the study of rauisuchians and early dinosaurs is important for our understanding of the ecosystem that developed in the rest of the Mesozoic. Using their bones, palaeontologists are trying to identify the reason why it was the dinosaurs that flourished so spectacularly across the Triassic–Jurassic boundary, and beyond.

Bodies were buried in the banks of the dry river bed near Qhemegha. Each mini-donga held its own collection, and a few of us were re-assigned from the main site to excavate them. For two days I had been focused on my section of the bank, sometimes alone and at other times working with Ralane to cut steadily but carefully into the slope following the line of bones, like a mole after earthworms. After a while the ache in my arm turned into muscular strength (promptly lost again after a few months back at my desk).

The rhythm of digging is mesmerising. Your whole world is dirt. Methodically peeling back each layer and searching, seeing the earth accumulating beneath you, is deeply satisfying. My eyes were dry and nostrils filled with dust. When I blew my nose, the hanky was Triassic brown.

In my reverie and focus I didn't notice the commotion of my colleagues at first. They were standing over to my left and exclaiming, bending over something uncovered at their feet. I put down my hammer and clambered across.

An enormous thorn lay half exposed in the mustard-coloured earth. It was a tooth, as big as my sun-browned thumb. We carefully removed more rock and found beneath it several more, sitting in place in an upper jaw bone the length of my forearm. A huge predator. The bones around it suggested we had an exceptionally complete animal. If this was part of the rauisuchian that Choiniere had identified earlier, it would be the most complete one known from South Africa.

The Karoo remains one of the best places in the world to find Triassic fossils. Along with Argentina and Arizona in the US, this place gives us the clearest insight into Triassic life on Earth. Taking stock, we had at Qhemegha representatives from the two most iconic Triassic lineages: early dinosaurs and rauisuchians. Unlike most previous finds we could gather detailed information on their location and position in the sequence of rocks, providing a high-resolution insight into the patterns of their evolution in this region.

The whole team was elated and couldn't wait to share the news with the rest of the village that night at the local inn. It justified all the hours and expense, and would bring income and tourism to the community over the coming years. We also hoped these bones could provide key new insights about evolution not just in South Africa, but around the Triassic world.

�†�†�†

From the Triassic onwards extrovert reptiles command a lot of attention – and superlatives. Giant killer crocodiles and elephantine proto-dinosaurs are quite distracting. They are like evolutionary fireworks, impressive to look at, extravagant, and exotic from our Anthropocene[*] standpoint. However, while reptiles were rising and exploding, the descendants of the mammal lineage were lighting their own revolutionary bonfires.

Like all animals, therapsids had been pummelled by the end-Permian mass extinction. Although some of them recovered exceptionally

---

[*] The current age, defined by the presence and impact of humankind.

quickly, they never regained the diversity of shape and form seen in the Permian. *Lystrosaurus* and the few closely related animals that lived alongside it were soon replaced by newer groups of dicynodonts which looked and behaved very similarly. These herbivores became the wildebeest of the Early to Mid-Triassic, constituting the most abundant plant-eating animals for over 20 million years.

Later in the Triassic, however, dicynodonts declined. By the Jurassic most herbivores larger than a dog were reptiles, and the only members of the mammal lineage left were cynodonts – including the ancestors of modern mammals.

But dicynodonts saved their best until last. A recent fossil skeleton from the Late Triassic of Poland, belonging to an animal the size of an elephant, shocked palaeontologists when it was revealed as not yet another early dinosaur, but an impressive *giant* dicynodont.

Called *Lisowicia*, this late survivor of the group may have reached seven tonnes, making it the biggest synapsid known until the mega-giants of recent geological history. It probably resembled an elephant in many ways, with a bulky, plant-digesting stomach, held aloft on straightened pillar legs. This limb shape has convergently evolved in multiple animal lineages as they've attained larger body masses. It reinforces the leg to support great weight, resulting in slower lumbering movement. At least three feet are always in contact with the ground when an elephant walks or runs.[*]

Lisowicia was elephantine, but it didn't have a trunk, and its canine tusks were reduced. It retained that oddball dicynodont beak, and was probably hairless. It would have been a fittingly impressive end to a group that had done so well in the face of disaster and usurpation by reptiles.

In the vast dry expanses of the Karoo, dicynodont fossils abound. Much of our first knowledge about them and other Triassic therapsids and cynodonts came from South African researchers – indeed they

---

[*] A feature often cited in the definition of 'running' is that all the feet must leave the ground together at some point, so you might argue that elephants can't actually run. However, it might be better to look at the stiffness of the legs and how they spring during movement, in which case elephants do a 'running walk'.

continue to lead such discoveries, often as part of international collaborations.

One key figure in South Africa's palaeontological history is yet another Scot,* who emigrated to the opposite end of the world at the end of the nineteenth century in search of adventure and opportunity. He found what he was looking for, and much more – the origins of humankind, and the beginnings of mammals themselves. Although his work was prolific in the story of scientific discovery, the lengths he went to in understanding the origins of humanity came at the expense of his own.

---

* We do get around, especially Highlanders, who were often packed off or encouraged to leave to form colonies throughout the British Empire so that their homeland in Scotland could be used to rear livestock and hunt game. Scots also participated in conquest, slavery and other colonial pursuits of the British Empire as soldiers, merchants and members of the landed gentry.

CHAPTER SEVEN
Milk Tooth

*Do not despise the creatures because they are minute ... And doubt not that in
these tiny creatures are mysteries more than we shall ever fathom ...*

Charles Kingsley, *Glaucus or the Wonders of the Shore*

Plaster of Paris quickly solidifies. As it does it tightens around your
skin like shrink-wrap, prickling as it tugs fine hairs. Our whole
team was albino to the elbows, as though the marble limbs of Greek
statues had been sewn onto our soft brown bodies. Snowflakes of
plaster splattered the plum-coloured rocks of the donga floor, and
splashes freckled our cheeks, some by accident, many from deliberate
playful flicking.

Job done and bones safely tucked in for the night, we were taking
turns to wash the plaster off before packing up for the day and heading
back to the inn in Qhemegha village. Tomorrow, Choiniere told me
as I scrubbed my skin russet using a scrap of hessian, I was to join a
small group prospecting a new location. We would all be looking for
fossils, but I was tasked with a special pursuit: 'I want you to find small
things.'

For all the years of collecting in South Africa and beyond,
palaeontologists predominantly gathered big stuff. Small fossils are the
most copious preserved organisms on Earth, but mostly in the form of
micro-organisms and invertebrates. Limestone, for example, is a cake
of ancient plankton. Other than the freakish giants of the Carboniferous,
arthropods are the gods of littleness, like tiny crustaceans or the
exemplary trilobites. The shells of molluscs litter the past like a
fisherman's midden.

For back-boned animals, though, most fossils collected are
reasonably large. This is not because small tetrapods didn't exist – far
from it – but their fragility and size means they were less likely to
persist in the fossil record, or be found by collectors. Bigger bones

over half a metre in length, as well as being more sought-after as objects to impress, are also easier to spot. In early collections emphasis was placed on finding and displaying large animals, especially their skulls. Scientists often knew species from their heads alone.

In the Triassic a curious thing happened to some therapsids that would have an impact on their fossil record and collection. Members of one lineage in particular became radically smaller.

The cynodonts emerged late in the Permian and made it through the mass extinction. These relatively small dog-like animals shared a number of characteristics with their closest therapsid relatives, the therocephalians, but in the Triassic they began to distinguish themselves. Those fenestrae in the skull expanded, and their cheekbones became deep and flared. A sagittal crest formed like a bone Mohawk on the skull roof. These features reflected the changing size and position of their jaw muscles, which were providing an increasingly precise bite.

Inside their mouths, the secondary palate had completely formed. Your palate is the bone at the top of your mouth, against which your tongue, at this moment, undoubtedly rests as you read this page. Run the tip of your tongue over it: can you feel the ridge down the middle and how the bone spans from left to right between your teeth, and backwards to the soft opening of your throat? If you were *Dimetrodon*, you wouldn't have this bone, you would feel a gap running front to back along the top of your mouth.* The left and right maxillae that hold your upper teeth didn't meet in the middle.

This lack of a hard secondary palate is the ancestral condition for all amniotes. Hard palates have evolved in multiple lineages of animals, including therapsids. It might seem like a trivial thing, but without it animals can't eat and breathe at the same time because the nasal cavity and mouth are not fully separated. You suffer the same effect when you try and eat dinner with a bad stuffy nose. The palate also provided a surface against which the tongue could manipulate food for chewing

---

* In truth, it is unlikely that if you were *Dimetrodon* you would have such a sensitive, mobile tongue. Our weird and versatile mouth muscles developed much later in mammals and were linked to food manipulation and milk drinking, as we now know from Jurassic fossils from China ... read on to find out more.

and swallowing. Lacking a hard palate poses a problem if you want to eat often, and even more so if you intend to chew your food. It is considered rude to chew with your mouth open, so the cynodonts, with their hard palates, were developing the first table manners.

In the rest of their bodies cynodonts were also setting themselves apart. Their shoulder blades changed shape as muscles altered, freeing them up for a greater range of movement. They had an increasingly upright posture, with legs more directly underneath the body. In their hips the pelvis and the top of the leg bones were also modified. A subtle rearrangement was occurring, one that would prepare the ground for the emergence of mammals.

A key shift in cynodont skeletons that took place through the Triassic was the regionalisation of the spine. In the human body, we take it for granted that we have different bones in our necks than at our waist. In most non-mammalian backboned animals the vertebrae are quite similar along the length of the body – at least by comparison to mammals. It can be dastardly hard to tell where a salamander backbone comes from for example, because although tail bones are usually quite distinct, from the third vertebra until you reach the pelvis they appear almost identical.

Humans, like most mammals, have four regions in our spines: the cervical (neck), thoracic (chest), lumbar (waist) and sacral (tailbone) regions. The bones in each region are starkly different, both in size and structure. This reflects different requirements placed on them, and is the result of an increasingly complex pattern in the spines of synapsids, therapsids and cynodonts, as recent research has shown.

In a common resident of the Triassic South African Karoo, we begin to see where our complicated mammal spines come from. One of our earliest cynodont relatives is a digging animal called *Thrinaxodon*. This is one of the first that really looks quite mammalian, with canine teeth, wide cheekbones, forward-facing eyes and a dog-like body that probably had at least some fur, if not a total covering. But more than this, it had a *waist*.

A punch to the gut and *Thrinaxodon* would have doubled over. This is because, like us, it had an almost rib-less lumbar region, the space between our chests and hips. Strange as it may seem, earlier synapsids had ribs all the way down. The bottom ones had reduced in the therapsids, an adaptation that was linked to their more active lifestyles. In *Thrinaxodon*, they were all but gone.

Having a waist means you can bend, imperative for mammals that must chase or outrun others. For those that run on four legs it leaves room for your knees and thighs when you stride, opening the door to new forms of locomotion like sprinting and climbing. It is thought that the loss of the lumbar ribs in mammals is also linked to the evolution of the diaphragm. This is the muscle that spans the base of your chest cavity and creates your in-breath as it contracts, like a biological bellows. Amphibians and reptiles have also evolved mechanisms to assist breathing, including snakes and lizards using their core body muscles to inflate and deflate their chests; expanding and contracting the throat (known as gular pumping) in varanid lizards; a similar system to a mammal diaphragm in crocodiles; and then some turtles, who have completely lost the plot and can breathe out of their bums.*

Recent research led by Katrina Jones at the University of Harvard[1] has shown that the increasing complexity of the mammal spine, from pelycosaurs to therapsids and then cynodonts, happened in what researchers call a 'stepwise' pattern. This means it wasn't gradual and steadily accumulating, but occurred at different times in different groups.

It appears that spinal regions, which came into sharp relief as cynodonts evolved through the Triassic, may have already existed in the underlying architecture of pelycosaurs. In previous research,[2] Jones and co-author Stephanie Pierce and colleagues had identified precursors to this regionalisation in earlier synapsids. They suggested that the reduced shoulder girdles of therapsids caused a change in their muscle arrangement, expanding some of the muscles of the shoulder and spine and driving changes in the vertebral column.

However it occurred, the spine that was present in the very first mammal predecessors in the Triassic was a gift that unwrapped a

---

* I said I wouldn't indulge in reptiles, but this fact was too ludicrous to overlook. Some water-dwellers have modified their cloaca – the shared hole for excreting and reproducing – to assist in gas exchange (that's oxygen gas, not the other kind). Sometimes nature is just ridiculous.

whole new spectrum of possible behaviours. Understanding how this happened is based in large part on Triassic fossils collected from South Africa in the early twentieth century, many of them by a man named Robert Broom.

There are some names running through scientific disciplines like seams of rock. They crop up repeatedly as you study a subject because they belong to people who contributed to the foundations of their discipline so significantly and prolifically that their work will always be remembered and cited. We've met many already: Buckland, Owen, Cuvier, the squabbling Cope and Marsh. Sharing the traits of being born male, white and usually wealthy, these men are heralded as figures of singular genius and determination. They are the ancient gods of palaeontology – science has built temples for them.

These foundational names are enduring, but a second look at the history of science shows it isn't ruled by academic gods, but by flawed men. Through combinations of privilege, luck, networking and personality, they often ascended on the backs of the poor, the female and the black.

The Scottish-South African doctor and palaeontologist Robert Broom is one such man. In the study of mammal predecessors, his surname is everywhere. His life's work is an important part of the story of mammal origins, but it also exposes raw wounds in human history.

Broom made discoveries illuminating animals we now know to be among our most ancient cousins. Their bones were disinterred from European colonies and shipped to institutions across Europe and the United States. In a post-Darwinian world, scientists were eager to fill in the gaps in the evolutionary story of life on Earth. They were searching for 'missing links'[*] – a term no longer used in science. Like most scientists at the turn of the nineteenth century, Broom believed evolution progressed from primitive to advanced, and the classification of living organisms was hierarchical. This hierarchy extended to human 'races'.

---

[*] As we saw in Chapter 2, this term is based on the idea of life being arranged in a hierarchical chain from one animal to the next. With modern understanding of evolution and taxonomy, it is now defunct.

Born in 1866, Broom was the son of a fabric designer and his wife. His father was a man of culture and learning, creating fabrics with the colourful swirls known as Paisley pattern, attending artisanal night school and enjoying arts and literature. Broom's mother was deeply religious. Broom took after his father's free-thinking attitude; always the contrarian, he joined the rebellion of evolutionists as a young man. But a maternal hand rested on his later beliefs, which were riddled with divinity and spiritualism.

It seemed unlikely that Broom was destined for global travel. He was a sullen-looking, sickly boy, suffering from recurrent bronchitis and infections. Trying to help him heal, the family sent him from industrial Paisley to live with his grandmother in the fresh air of the Scottish coast. There, the gift of a microscope from a local naturalist – and the training to use it – ignited his interest in the natural world. Despite only sporadic schooling he was sharp and inquisitive. Broom began seeking out the countryside and studying the animals that lived there.

Like many young naturalists, Broom was a collector. His cousin, an engineer stationed in China, sent home boxes of curiosities for his growing collections, including pickled snakes and human skulls. There is no record of where these items came from or how they were acquired, but given the time period it is unlikely to be a wholesome source.[*] Broom discovered his first fossils at age 15, at a limestone quarry during one of his long countryside walks. The ancient shells and corals would later be the basis for some of his first scientific papers.

He studied to become an obstetrician (at that time known simply as a midwife) at the University of Glasgow, where his difficult, solitary personality caused problems with tutors. Graduating at the top of his class, Broom worked at Glasgow Maternity Hospital, but was distracted with wanderlust. He travelled first to the United States, before moving to Australia, where he collected plants and animals and saw his first platypus, indulging in the life of a naturalist. Increasingly fascinated with marsupials and monotremes, he studied opossums, the same creatures once thought to have been found in the Jurassic rocks

---

[*] Interest in human skulls in particular was tied to scientific racism, as well as their being sought after as colonial 'trophies'.

of Stonesfield in England. Intrigued by what Australian animals could reveal about mammal origins, Broom began meticulous collecting and study to understand their anatomy and evolution.

In 1896 Broom made plans to emigrate to South Africa and continue his search for mammal origins. He familiarised himself with the Karoo fossils held at the Natural History Museum in London. Their initial cantankerous describer, Richard Owen, had just died. He and his contemporary in the United States, Edward Drinker Cope, had already realised that therapsids were key to understanding where mammals came from. Broom knew that if he was to uncover their origins, he had to take a closer look at their fossils.

Owen and Broom had corresponded before his death, and the old man thought highly of him. The specimens Owen described had been sent to him by collectors from South Africa seeking an expert opinion – which he readily gave – and the fossils found 'safe-keeping' in the British Museum of Natural History (now the Natural History Museum in London).

But in the 1880s another prestigious palaeontologist had taken charge of studying the Karoo specimens. Harry Govier Seeley not only studied them, but travelled to the southern tip of Africa where he toured inland and made extensive new finds with his white South African collaborators. These fossils were then usually removed from their country of origin. In particular Seeley was responsible for collecting and describing Permian-aged therapsids and exceptional pareiasaurs, but he also described Triassic creatures that lay at the very base of the branch that produced our first mammal kin, the cynodonts.

By the time Broom was making his plans it was generally accepted, thanks to the study of South African fossils, that mammals probably came from among the therapsids.* But which group, or groups, were their ancestors remained a mystery – one Broom intended to solve.

---

* There were some researchers who believed for a time that they might have evolved from amphibians. Although this was later disproved, the common misconception that reptiles evolved from amphibians, and mammals from reptiles, persists. As we saw in Chapter 3, all three groups actually evolved from a tetrapod ancestor that was the precursor to all of them.

Seeley was not keen to help Broom in his southern venture. Behind the scenes he machinated to prevent this upstart Scot from encroaching on his research. He asked his South African contacts to make sure all future fossil discoveries were sent to London and shown to no one else.

Broom and his wife Mary arrived in Cape Town on the SS *Goth* in 1897. He found the museum, at Seeley's bidding, unwilling to allow him access to their collections. He had no contacts and had to take locum medical jobs as they came. Finally he was offered a position in Port Elizabeth, a seashore town in the Eastern Cape. Broom once again immersed himself in natural science and was even allowed access to the Karoo fossils in the town's museum – the first he'd seen in their country of origin. But the medical workload interfered with his studies in Port Elizabeth, and soon the Brooms headed inland to devote more time to prospecting.

With encouragement from friends, Broom resolved to break Seeley's monopoly on the fossils of South Africa. He spent every spare moment scouring the dongas, collecting new fossils that he and his colleagues could examine without interference. He struck up friendships with white South African collectors, some of whom were sore from the dismissive treatment they'd received from British scientists, who often took for granted the precious specimens they received from enthusiastic amateur fossil hunters. Broom, however, appreciated their hard work, and added his own.

In 1903 Broom was appointed Professor of Zoology and Geology at Victoria College in Stellenbosch. He was quite different from the previous lecturers of the institution, refusing to treat the students 'like big schoolboys'[3] by doing roll-calls or drilling them for examinations. Instead he encouraged them to think for themselves, enticing them with unrehearsed lectures that often meandered and digressed off-topic into politics and religion. Having an obstinate, self-reliant personality and lack of talent for political manoeuvring meant Broom's time at Stellenbosch didn't last, and he struggled to find the scientific positions he believed himself entitled to. He eventually left for the mining town of Springs, not far from modern-day Johannesburg.

From there Broom focused on collecting from the Karoo, at liberty to describe material that interested him. He continued to misstep as he ignored authority in the hunt for new material, but still hoped to be noticed for his substantial contributions. Broom became the centre of competition for fossils between museums in South Africa, Britain and the United States, dispersing his collections to institutions like the University of the Witwatersrand and the American Museum of Natural History, where he struck up a close collaboration and friendship with Henry Fairfield Osborn.

What distinguished Broom's research on the origin of mammals was his interest in moving beyond simple description. He wanted to grasp family relationships, and asked detailed anatomical questions, focusing on the kind of characters of the skeleton that we still use to distinguish different animal groups. We met some of these traits in earlier chapters: synapomorphies like the placement of certain bones, or the development of differentiated teeth. These were the foundations of understanding relationships among early synapsids, therapsids and cynodonts, and are now used in a form of analysis called phylogenetics.

Phylogenetics, simply put, is the study of evolutionary relationships. It is now carried out statistically, using computer algorithms to assemble the most likely family tree based on inference. The information used in these computer programs varies from DNA sequences to characters of the skeleton. Although it can't give you *the* answers, it can provide the most probable hypotheses, based on statistical likelihood.

For fossils of course, the skeleton is usually the only source of information. Palaeontologists note the features of the skeleton and give them 'scores': for example, scoring a 0 if a feature is not present or a 1 if it is. As well as presence and absence, complexity is added to these scores for features that can have multiple 'states'. So a tooth might have a single pointed cusp (scored 0), two cusps (scored 1) or three cusps (scored 2). All of the scores are placed in a large table called a character matrix, and using this, the phylogenetic analysis works out how animals are related.

Understanding family trees through phylogenetics now requires coding programmes and lots of maths, but it works based on basic foundations established over the last 250 years. Closely related animals share more features than distant relatives, and so it is possible to assemble a family tree by observing these similarities. Although our understanding of relationships based on similarity has been complicated, refined and added to in the last 80 years, this foundational idea was already well understood by Robert Broom and his contemporaries.

Broom used this approach to recognise distinguishing features in 369 new types of therapsids, of which 210 remain valid to this day. It was he who named *Moschops*, the TV star therapsid we met in Chapter 5. He applied his knowledge to the pelycosaurs of the United States – including those of the Texan red beds – comparing them to therapsids and recognising the close affinities of the two groups. In 1937, Broom found one of the first members of the group of therapsids to which we belong – the cynodonts – which he called *Procynosuchus*.

Broom didn't confine his observations of anatomy and skeletons to the Triassic fossil record. His career took a turn in the 1930s when he discovered early hominin fossils in caves near Johannesburg (when asked about the discoveries, Broom – then in his seventies and increasingly aware of his own antiquity – indulged in his growing superstitious outlook, claiming to be guided to the fossils by spirits). It was this find, rather than his work on mammal predecessors, that made his name world-wide.

One of the caves, called Swartkrans, is now part of the Cradle of Humankind World Heritage Site, and has yielded bones from the last few million years that include *Homo ergaster*, *Paranthropus* and *Homo habilis*, all members of the bushy side-shoot of apes that includes modern humans. Some of the earliest evidence of the use of fire and bone tools is known from the site.

Through this, and many other fossil discoveries, Broom played a key role in science's quest to understand our human origins. But the part of this quest that people don't like to talk about is the unsavoury beliefs held by many of those who pursued it. Their collections usually came at terrible human cost.

Broom and his contemporaries not only categorised fossil skeletons, but also the bodies of indigenous South Africans. The products of their work were sent to institutions across Europe and North America to be studied and arranged. Western scientists wanted to uncover the 'missing link' between apes and the kind of humans that they placed at the top of nature's hierarchy, closest to God: themselves. Everything else was just another 'specimen'.

In the nineteenth century there was a widespread belief that the progenitors of the human race were European. This view was held by some of the leading scientific figures, including Georges Cuvier. It was thought that all people originated from these European ancestors after the flood, and that they had since 'degenerated' into the different races of the world. Pro-slavery campaigners even argued that slavery was good for non-Europeans, because it saved them from inevitable extinction as 'doomed' races.

By the end of the nineteenth century, however, scientific theories about the origins of humans had changed. Charles Darwin was among the first to argue that humans originated in Africa, an idea not widely accepted until the latter half of the twentieth century. But with this new idea in mind, researchers turned their attention to the vast southern continent for answers.

When he began work in South Africa, Broom commented that 'the most interesting specimens were the natives'.[4] With the mechanism of evolution now generally accepted, many of those searching for our human origins – including Broom and contemporaries such as Osborn – believed that indigenous Africans represented the so-called 'missing link' between apes and humans. Western scientists thought that by studying the physical traits of African people, they could illuminate the origins of man.[*] The skeletons of indigenous people from parts of Africa as well as Australia, Indonesia and the Americas were sought after for study, and obtained by disturbing means.

Collectors in South Africa regularly shot people to satisfy the demand from museums. Although it is unclear if he murdered anyone outright, like others Broom dug up the graves of indigenous South Africans, hoping to amass 'one of the best collections in existence'.[5] He collected dead bodies from the local jail, burying them in his garden to dig up later, or butchering them with kitchen knives and boiling the heads on the stove to remove the flesh.

External features were also studied. Palaeoanthropologists (scientists who study human fossils) began collecting plaster casts of the faces of

---

[*] Christa Kuljian's book, *Darwin's Hunch*, documents the subject of science, race and the search for human origins in South Africa and beyond. I highly recommend it for a fuller understanding of this subject.

people from around the world, believing they could trace the evolution of their features and categorise them. The plaster cast took time to dry and often hurt, pulling out facial hair as it was removed. Researchers sometimes lied to the people they cast in order to persuade them to participate, saying the process had healing benefits. Hundreds of these face masks, along with measurements of body parts (including female labia) were added to the amassed skeletal collections of museums and universities. The 'gallery of faces' on display in the Medical School at the University of the Witwatersrand was once described as the 'world's finest collection'.[6]

Addressing the racist history of collected objects in museums is known as decolonisation. Calls to decolonise collections are increasingly urgent. The process includes returning the context of collected objects: their origin, who collected them, how they did it and why. Such context was regularly stripped from objects, deliberately misrepresented or ignored. Without this background, the people who suffered are written out of history.

Our museums are still arranged from a white Western perspective, with items donated by wealthy patrons and collected (we can usually assume stolen) by those who benefited from empire. In efforts to decolonise, objects may have their display information updated, while others are repatriated. Difficult conversations are taking place as staff and audiences struggle to figure out what museums are for, and who they are intended to serve.

In many institutions, there are objects tucked shamefully out of sight. Hard to face, they are locked away in cupboards rarely opened, in dark storeroom corners. Researchers stumble on them in the course of their studies, balking at the yellowed labels. It was just such a specimen, a human skeleton 'collected' by Robert Broom, that first alerted me to his activities in the pursuit of 'specimens'.[*]

---

[*] Sadly, I don't know who the skeleton belonged to. It was seen by a colleague during a visit to a collection in South Africa. He asked about it and was told (briefly) about Broom's collecting history. This colleague mentioned it to me when I talked about Broom's role in mammal research for my book. I investigated further, but so far have no information on the location of the original skeleton and who it belonged to.

We'd like to think science is impartial and that the work of researchers can be separated from their beliefs and practices. But structural racism and colonialism remains tightly stitched into the fabric of science. Broom's discoveries may form the basis of much of our knowledge of Permian and Triassic mammal evolution, but they can never be removed from the socio-political context of the man, the country and the wider world we share. We are all still affected by this legacy.

As we walked further from Qhemegha village and the dig site, I felt smaller and smaller. The earth of the veldt was drought beige except for sporadic sprinkles of buttery wild flowers. My nose was bleeding again from the dry heat, and my boots rasped through the dry shrubs like a rhythmic cough. My companions were spread out, each surveying their own path across the plain. With each step we drew closer to the distant ridge. Passing a shepherd I waved and complimented him on his handsome sheep. With their long legs and short wool, they looked more like goats to me than the walking clouds I'm used to seeing at home.

Reaching the ridge our team found a place where the underlying rock layers were exposed. I dropped to my hands and knees. To find small things you have to get down to the ground. As so often happens to those in pursuit of the small, I was so busy picking over the rocks for evidence of tiny bones that I completely missed the huge row of vertebrae the others were admiring, each one as big as a soup bowl. For me, the devil was in the detail.

Soon I found something: white wisps in the purple stone, faded scribbles. Were they bones? I pulled my hand-lens out from where it rested down my shirt front. The metal was hot against my eyebrow. They were broken, but the magnified shapes looked complex in cross-section, like jumbles of tiny skeleton, only millimetres in length. I was excited. I began gathering the pieces in a plastic bag, like Scrabble tiles. I separated especially promising ones and kept them aside to show the others. Tiny teeth or skull fragments maybe? When things are so little it's hard to tell without a microscope.

After an hour there were two piles of bones ready for collection from our new site. One was a cairn of rock footballs that each held large fossils, enough to fill a bathtub. And beside them sat my two little sachets of gravel, the contents of a bag of crisps at best.

The rest of the team squinted at my finds with bafflement, but I was pleased. If even just a few of these specks turned out to be the bones of small mammals – or other tiny animals like amphibians or lizards – they'd be some of the only ones known from the Triassic of South Africa. Where it took a truckload of bones to glean information from just one dinosaur, for these microbeasts I could fit decades of transformative scientific research into my pocket, and still have room for a cereal bar.

How and why did the mammal lineage reduce so radically in size? Many of their Permian predecessors were huge. Those that pulled through the end-Permian mass extinction were reduced in size, but certainly hefty. As the Triassic continued, from the precursors like Broom's *Procynosuchus* and Seeley's *Thrinaxodon*, the cynodonts split into multiple groups. There were carnivorous ones, like cynognathids, probaignognathids and trithelodontids. Plumping for a life of omnivory and herbivory were groups like the diademodontids and tritylodontids. Together these animals are all known by researchers as non-mammalian cynodonts. This differentiates them all from the final group to appear in the Triassic: the mammals, or more strictly speaking, mammaliaforms (we'll come back to the difference between those two names in a moment).

How these different cynodont groups are related to one another continues to be debated. Different phylogenetic analyses spit out differing results with each new input of information and interpretation. A lot of them are superficially rather samey to look at (although specialists will berate me for saying so). Most resembled *Thrinaxodon* in body plan, being ferret-sized to pig-sized with a few variations in their skulls or teeth.

In the line-up of non-mammalian cynodonts there are a few groups that most researchers agree are the likely candidates for the closest relatives of mammals: the carnivorous trithelodontids and Brazilian brasilodontids (the current favourites), and the herbivorous tritylodontids. Settling which one is closest is a question that has preoccupied palaeontologists for decades. It is unlikely to ever be resolved to complete satisfaction.

One thing that everyone *can* agree on, is that the first Triassic mammals were very small. In modern zoological studies small-bodied animals are often defined as being under 5kg (11lb) in body mass, which is smaller than a fox. In the Late Triassic the first mammals shrank well below this threshold, with none more than a few hundred grams in weight.

Being small meant that these mammals were often overlooked during fossil-hunting expeditions. By the start of the nineteenth century the first Mesozoic mammals had been found in the Jurassic rocks of England. But it was well into the twentieth century – and with 100 years of Karoo fossils under the belt – before the first Mesozoic mammals were found in South Africa. They came from the Late Triassic, and together with discoveries in Britain, would prove to be some of the most important fossils in the world for understanding the origin of mammals.

Not far from Qhemegha, across the border in Lesotho, the first Late Triassic mammals from southern Africa were discovered. The first was from Mafeteng, and it was an added extra, nestled in the rock matrix surrounding a new species of dinosaur. It was spotted during preparation of the reptile in 1962, and comprised a single jaw bone similar to the find from Stonesfield. It was named *Erythrotherium* by South African palaeontologist Alfred 'Fuzz' Crompton, another one of the big names in mammal palaeontology since the latter half of the twentieth century.

The second Triassic mammal discovery was not made by chance. It happened on a joint field trip of the South African Museum, Yale University, British Museum and London University in 1966, including Crompton and fellow palaeontologist Farish Jenkins, another leading researcher in mammal palaeontology. Among their field crew was a young research assistant at the South African Museum, a woman named Ione Rudner. Small and slender, with a long pony-tail swept over one shoulder,[6] Rudner seemed an incongruous addition to a team of stereotypical outdoorsmen. She often found herself expected to cook meals and make sandwiches, despite her expertise as a palaeontologist and archaeologist. Newspapers reporting on the subsequent discoveries of their field expedition referred to her as a 'housewife'.[7]

One day while the other researchers were collecting a dinosaur fossil, Rudner went prospecting along a goat track a few kilometres east of Fort Hartley. Passing a ledge by the path she noticed a pale 'mosaic of bone' in the rock. This turned out to be the crushed skull and lower jaw of a new species of Late Triassic mammal, no bigger than a mouse.[8] It was named in her honour: *Megazostrodon rudnerae*.* A second skull with postcrania (the body of the skeleton) was also found from a site to the north in Clocolan by a different researcher named James Kitching. Together, these discoveries made *Megazostrodon* one of the best known Late Triassic mammals in the world.

There are only a handful of fossils of Late Triassic mammals. That isn't just phraseology, I mean it literally: you could probably hold every one of their tiny bones in your open palms. As well as the two from South Africa there are a scattering of fragments in Europe, including the most famous of all, a Welsh beastie named *Morganucodon* ('Glamorgan tooth', named for the region in which it was discovered). This little celebrity lends its name to the group of early mammals called morganucodontans. Also from Wales comes *Kuehneotherium*, and from China, *Sinoconodon*.

None of these animals was bigger than a mouse, but they are the ancestors of us all (speaking poetically – we don't know which ones, if any, were our direct ancestors and we never will). This smattering of fossils, discovered through the 1960s, suddenly brought our Late Triassic emergence into sharp focus. Until then there had been a conspicuous gap between the many cynodonts of the Triassic and the opossum-like animals of the Jurassic. The new tiny fossils plugged it, charting mammal emergence from amidst the fray of non-mammalian cynodonts.

The first mammals are distinguished from the rest of their relatives by one feature in particular: their jaw joints. In the absence of soft tissues and faced with a continuum of creatures assembling mammal-like traits in fits and starts, researchers had to draw a skeletal line under things. There had to be a moment in the fossil record they could point to; an X-marks-the-spot of mammal origins. The one they chose was the jaw joint. If the back of the jaw forms a joint between the dentary bone and the squamosal bone (part of the cheek

---

* *Megazostrodon* refers to the tooth shape, 'big girdle tooth', and *rudnerae* to Rudner.

Above: Dinosaur footprint (foreground) on the shoreline of Trotternish, Isle of Skye, Scotland.

Above: Fossil skeleton of the pelycosaur, *Edaphosaurus*.

Below: Cast of the skull of the therapsid, *Estemmenosuchus*.

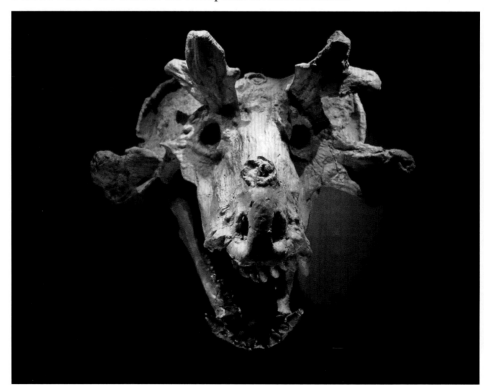

Above: Fossil skeleton of the pelycosaur, *Dimetrodon*.

TELEOSAURUS

DIMORPHODON

DUCK-BILL

Right: *Teleosaurus, Dimorphodon* and two 'duck-bill' platypuses, illustrated in *Nebula to Man* (Knipe, 1905).

Top: Zofia Kielan-Jaworowska, world-leading Mesozoic mammal researcher, on fieldwork in Mongolia.

Centre: From left to right, Zofia Kielan-Jaworowska, Zorikht, Teresa Maryańska and Gunzhid at Altan Ula camp, Mongolia.

Bottom: From left to right, Zhe-Xi Luo, Zofia Kielan-Jaworowska and Richard Cifelli at Kielan-Jaworowska's home.

Right: *Eomalia*, member of the mammal group Eutheria, found in the Jehol Biota of China.

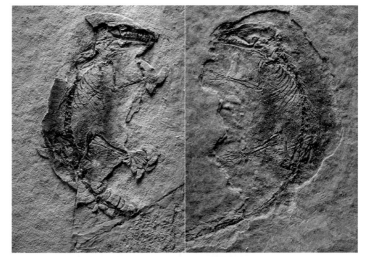

Right: Jurassic docodontan, *Docofossor brachydactylus*; the first mole-like digging mammal.

Far right: *Maiopatagium furculiferum*, one of the first gliding mammals, from the Jurassic.

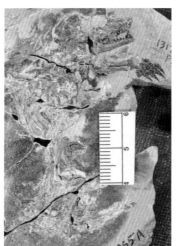

Above: The first mammal fossil ever described from the Mesozoic in 1824: *Amphitherium prevostii.*

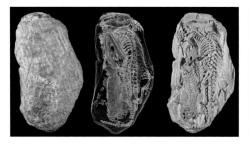

Above left: Fossils of the cynodont *Thrinaxodon*, and bedfellow *Broomistega*, digitally reconstructed using X-ray synchrotron scans.

Above right: Maxillary canal (green) and sinus (purple) in *Thrinaxodon*.

Left: William Johnson Sollas's serial grinding machine, precursor to X-ray CT scanning.

Below: Digital reconstruction from X-ray CT scans of the ear bone (petrosal) of the Middle Jurassic mammal *Borealestes*, from Scotland.

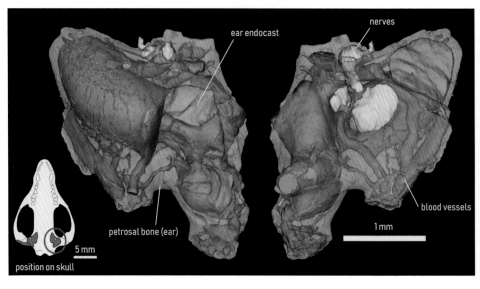

ear endocast

nerves

petrosal bone (ear)

blood vessels

1 mm

5 mm

position on skull

Above: Digital artistic reconstruction of *Borealestes serendipitus* (left) and
B. cuillinensis (right) skulls, Middle Jurassic mammals from the Isle of Skye,
Scotland.

Right: The Jurassic
docodontan mammal
Agilodocodon scansorius;
an early tree-climbing
specialist.

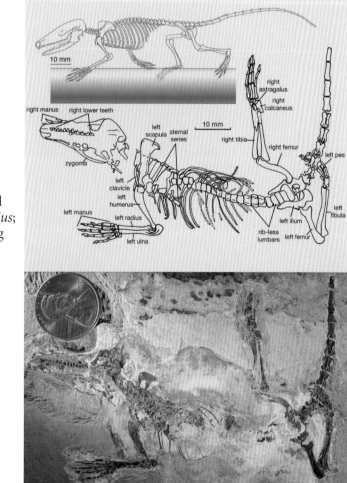

Right: Jaw of the mammal, *Boreal-estes*, moments after its discovery in the Middle Jurassic rocks on the Isle of Skye, Scotland.

Right: One of the first fossil discoveries on the Isle of Skye, in 1971, a limb bone no bigger than a matchstick.

Below: Robert Savage on fieldwork on the Isle of Skye, Scotland, 1973.

area in the skull), hey presto, it's a mammal! If there is no contact between these bones, then there is no mammal.

In truth, even this delineation has a caveat. Late Triassic mammals were cheating. They had their brand new dentary-squamosal jaw joints, but still retained the remnants of the non-mammalian cynodont joint too. These were formed between the much-reduced articular bone still clinging on at the back of the jaw, and what remained of the quadrate, a tiny skullbone tucked below the cheek. But they are undoubtedly the very earliest of our kin, and the precursors of the next 220 million years of hot furry innovation.

Their emergence took place from the Late Triassic to the earliest days of the Jurassic. It straddled the end-Triassic mass extinction that re-arranged the great reptile lineages, demoting crocs and putting dinosaurs in the prime slots of Earth's terrestrial ecosystems. Almost all non-mammalian cynodonts died out – only the hardy tritylodontids and scatter of trithelodontids remained – and reptile equivalents took most of their positions. Dinosaurs burst at the seams, not only occupying all of the ecological space that therapsids and crocodile-line reptiles had previously worked in, but magnifying to become giants of the Earth – and of the human imagination.

At a glance it might seem that the first mammals were relegated to puniness. They apparently lost the battle against their reptile overlords, the so-called kings of tyranny. Dinosaurs are rendered roaring from a thousand book covers, stomping and shaking our screens, behaving altogether more like monsters from a nightmare than living creatures. Mammals, meanwhile, are nothing but their cowering snack food.

In 1925 the mammal palaeontologist George Gaylord Simpson wrote an updated version of Richard Owen's great monograph on Mesozoic mammals. Despite making his career through their study, Simpson wrote to his sister, 'I've belaboured the filthy beasts for about four years now and I hope to heaven no one finds any more during my lifetime.'[9] They remained the lowest form of sucklers, even to him. Few people were impressed with their minuscule bodies.

This negative attitude has not really changed. From the many books and papers to popular science pieces and public talks, Mesozoic mammals are maligned through language, art and simile. They lurk, creep and scuttle in the shadows of noble reptile brethren. They are verminous and 'rat-like'. They are oppressed by the mighty, subjugated. They are the butt of jokes. They are losers.

Mammals, we are told, did nothing during the time of dinosaurs, so we skip over them in our narrative of life. Most books on mammal evolution only begin when they are finally free of their masters, and get bigger. Because bigger is always better. Right?

I'd like to propose a drinking game. Next time you watch a documentary about wildlife, fossils or evolution, take a drink every time the narrator says an animal is *successful*. If they say something is the *most* successful, down your glassful. Be prepared to get tipsy, because success is one of the most overused descriptions in nature documentary film-making – and perhaps the whole of zoological science.

What exactly is success? Colloquially, it's a term often applied to big organisms because we consider size analogous to success, whether implicitly or explicitly. Whales and elephants, for example: they must be doing well to grow so large. Speaking scientifically, success is somewhat analogous to fitness. Usually this is defined in terms of passing on genes: the better you are at getting your DNA into the next generation, the fitter you are in the evolutionary sense.

But when most people use the word success about animals – either colloquially or among scientists – I would argue they are rarely referring to an exacting biological definition. They are usually thinking of one or more of four things: persistence through geological time; abundance or diversity; wide geographical range; and a vague perceived notion of being 'better' – most often associated with apex predators, which for some reason we think are doing a better job at existing than their far more speciose herbivorous prey. Sharks are a good example of what most people would say was a 'successful' group, and I agree with that.

Whether you subscribe to the biological definition or the colloquial one, one thing is undeniable: small mammals are *successful*. So are insects, lizards, fungi – even bacteria are incredibly *successful*. And yet we rarely discuss these organisms in these terms.

It is time to rethink our idea of what it means to do well in the game of evolution.

Mesozoic mammals, it's true, reduced their body mass in the Late Triassic. They ended up much smaller than their predecessors, whereas the dinosaurs grew larger. Far from being a defect, however, mammal smallness was a *revelation*. They pioneered a new kind of terrestrial vertebrate miniaturisation. With no space among the giants, they took a different route: they perfected being tiny.

The faster metabolism and warmer blood of mammals were a gift from their predecessors. These traits meant they could not only become smaller in the Late Triassic, but also nocturnal. The coldness of night was no barrier for an animal carrying its own heating system. We know the first mammals were nocturnal thanks to the legacy left in the eyes of every mammal on Earth today.

Inside our eyes are two main types of photoreceptor, or light-sensitive cells. The first, called cones, are less sensitive to light, but provide higher resolution. This is better for daytime vision. The other receptors, called rods, are more sensitive to light, but this sensitivity comes at the expense of resolution. This makes them much more useful during low light conditions.

Surveying mammals alive today, biologists noticed that they have lots of rods in their eyes, but only a few cones (of the four types of cone present in other vertebrates, mammals only have two). This gives them good low-light vision, but as a result almost all mammals are colour-blind – they can't distinguish red-yellow-green parts of the light spectrum. Primates like ourselves are among the only mammals that have re-acquired good colour perception, a feat achieved through mutations in our genes.* This primate technicolour is by far the exception and not the rule, probably selected-for owing to the advantage it gave our common ancestors foraging for ripe fruits and young fresh foliage.

---

* Once lost, the original cones and opsins (proteins that form visual pigments) can't be reclaimed. Evolution can only work with what exists, it can't go backwards. Instead, mutations in the existing opsins have led to primates re-acquiring trichromatic colour vision in their two types of cone. It is such a useful mutation, it happened more than once: at the base of the primate tree in the Old World (including macaques, gorillas, orangutans, chimpanzees and us), in the howler monkeys of the New World, and in the honey possum (*Tarsipes*). But our colour vision is still a far cry from the glorious visual world of birds.

The widespread abundance of rods and paucity of cones in mammal eyes today is testament to an enduring night-time legacy. From at least 220 million years ago, the ancestors of all living mammals exploited the darkness. This has been named 'the nocturnal bottleneck' – a rather negative-sounding name for something quite extraordinary. Some scientists suggest this mammal night-time activity pattern could have lasted the entire Mesozoic, but there is no way to know for sure how varied their activity patterns really were; some species may have returned to being active during the day.

In the modern world most mammals are still at their most active at night. Breakthroughs in night-vision camera technology are revealing their exploits, including species we assumed were daytime creatures. Cheetahs, for example, have been shown to hunt on moonlit nights, while rhinos meeting around watering holes exhibit exquisite social behaviour not observed under the sun. Small mammals still prefer dusk and darkness for the advantages they provide in avoiding predators – after all, they are quite literally bite-sized.

Although it undoubtedly shaped the first true mammals, for synapsids as a whole being active at night was not a Triassic invention. In a study in 2014[10] the palaeontologist Kenneth Angielczyk and colleague Lars Schmitz surveyed synapsids from as far back as the Carboniferous, looking for signs of nocturnality. They examined the dimensions of the eye orbit, and in earlier synapsids, the sclerotic ring inside the eye, both of which have been shown to link to eyeball size and visual ability in living animals. Their results suggested that some of the very first synapsids over 300 million years ago could have been active at night. This suggests that being active at night was a way of life that emerged several times in the history of mammals, all the way back to the Carboniferous.

Although it was not the first time synapsids had turned to the path of darkness, the nocturnality of Triassic mammals was wholesale and transformative. Even their closest relatives such as the non-mammalian tritylodontids were exploiting low-light lifestyles. This dedication to the dark had far-reaching consequences for mammal senses.

Rods in the eyes made use of what little light was available, but the world without light meant mammals had to rely on other means of finding food and mates. They already possessed fur and whiskers by this time, probably evolving in their predecessors as an adaptation for navigating burrows (more on the evidence for the development of fur later). In the lightless world below ground, these delicate structures would have enhanced the tactile feedback received by the skin, making it easier to feel their way through the darkness. Such supersensitivity to touch would have proved equally vital in their nocturnal world among the Mesozoic foliage.

Scent and sound, now two of the most important senses used by mammals for communication and hunting, were enhanced in the Late Triassic. Although most animals can smell and hear, mammals are among the smelliest and noisiest. Observing their brain anatomy from the few complete skulls in the fossil record, there is a clear increase in the parts of the brain that process the senses, primarily the olfactory bulbs at the very front. Evidence for this came from the very smallest of fossil mammals, *Hadrocodium*, which weighed less than a sugar cube. It was found in the Lufeng Formation, an Early Jurassic rock layer in Yunnan Province in southern China. Its description in 2001 was accompanied by an iconic artistic reconstruction of the animal crouching next to a nearly equal-sized paperclip. Its skull was smaller than a grape, but it had a large brain compared to body mass (*Hadrocodium* means 'large head').

Hadrocodium was extraordinary for its smallness, further cementing this as the stereotype for Mesozoic mammals – but it told us so much more. The brain case was more or less intact, allowing researchers to reconstruct the shape of the brain itself. As well as being large overall, it had big olfactory bulbs at the front. These bulbs are the place where the brain makes sense of signals received from the nose, and they are larger in animals that rely on smell. We therefore know that by *Hadrocodium*'s time, mammals were sniffing their way to success.

Mammal hearing is also a super-sense. There are species that detect sound at super-high and low frequencies, from the tiny conversations of rodents and pinpoint moth-detection of bats, to belly rumbles between savannah elephants and the ocean-spanning songs of lonely whales. None of this would be possible without the refinement of the mammal·ear, a process we now know could only take place thanks to the miniaturisation of our Triassic ancestors.

The unique ear anatomy of mammals includes bones not present in other tetrapods: the malleus and incus in the middle ear. These work with the stapes to enhance sound, widening the range of detectable frequencies. We know from the fossil record that these bones were originally part of the jaw. One of them is the articular, which once formed part of their cynodont ancestor's jaw joints, and still comprises part of the jaw joint in other vertebrates. Through mammal evolution these bones shrank and ended up incorporated into the middle ear. What we didn't know until recently was how those jaw bones could have made the leap to their new home within the skull.

Recent research from palaeontologists in Great Britain and the United States suggests that it was being small that allowed mammal jaws to transform. By analysing the jaws of non-mammalian cynodonts and early mammals, a team led by Stephan Lautenschlager discovered that having a smaller jaw (due to being smaller overall) meant the first mammals could bite harder while reducing the stresses placed on their jaw joints.[11] This would have changed the way they bit, which had repercussions for the position of jaw muscles, freeing up the bones at the back of the jaw. These were then available for use in refined hearing.

Another unexpected repercussion of their physical reduction was that it turned mammals into little nippers. In their mouths Triassic mammals were equipped with the latest in weaponry. If you look at the gnashers of a crocodile you can see that they are in different stages of eruption from the gumline: some are fully out, others are still half-emerged. This is because they are being replaced throughout their lives. When one falls out or is broken, the next pushes up to take its place. Crocodiles have similarly shaped teeth all along the snout, and their irregularity of eruption doesn't matter because all they are for is getting a good grip on food before pulling and swallowing.

For modern mammals like the hedgehog, there is just one main replacement event: the loss of baby (deciduous) teeth and growing in of the permanent adult ones. We call this diphyodonty. It is a hallmark of all mammals, occurring in all modern toothed mammals with only slight variations in the pattern and timing. The hedgehog also has incisors, canines, premolars and molars, a differentiation in tooth types that began in the therapsids and then continued in cynodonts. It allowed animals to start feeding differently: canines can puncture, incisors can grip, specialised molars can shear, slice and grind.

Permian and early Triassic animals still replaced their teeth throughout their lives. But as the Triassic continued, this pattern changed for some groups. The gomphodonts were among the first to reduce their replacement rates. They might sound like a type of troll from a fantasy novel, but are actually non-mammalian cynodonts, first named by Seeley from his studies on South African fossils. Although his original definition of them has been modified since then, he noted early on that these animals had tooth replacement patterns unlike their therapsid predecessors. Their postcanines were replaced in groups, with new ones emerging at the back. The timing appeared to allow better occlusion between the upper and lower teeth, improving the ability to process food by chewing.

But an altogether more radical change emerged in the Late Triassic mammals. *Sinoconodon* was the last of its kind to replace its teeth more often. The rest of the lineage reduced their pattern and became diphyodont.

This single replacement is significant for two reasons, one of which ties in to the very core of what it is to be a mammal. Deciduous baby teeth are also known as 'milk teeth' because their loss often coincides with weaning. Mammals have taken the unique approach of feeding their young on milk until they reach maturity. We can assume therefore, that if an animal was diphyodont, it fed its young on milk.

Secondly, if you want to take full advantage of specialised, complex-shaped cheek teeth to slice, shear and grind your food, those teeth have to occlude precisely. Scissors don't cut paper if their edges don't meet. You can't grind spices if your pestle is too big for your mortar. By replacing their teeth only once, mammals can ensure that they remain a snug fit throughout their adult lives.

With this dental yin and yang secured, mammals began to experiment with tooth shape and new diets. A world of dietary innovation was theirs to explore.

It can't be ignored that small mammals hold a particular position in the food chain. The abundance that marks their success also makes them a good staple for a predator's diet. As in modern ecosystems, Mesozoic mammals would have been on the menu for larger animals – who isn't? But as anyone who has ever tried to catch a mouse or escaped pet gerbil can attest, when you grab a handful of unwilling floof you often get more than you bargained for. Their bite may be small, but it really hurts, like stepping on Lego. To compensate for size many diminutive mammals are ferocious. You can imagine Mesozoic mammals would have done some serious damage with their new set of permanent needle teeth.

There was, however, an even greater danger to any reptiles that tried to snatch a mouthful of Triassic mammal. They could have been stung.

Like the modern platypus, it is thought that the first mammals might have possessed a venom-laden spur on their ankle. In the platypus it is positioned just behind the foot and is shaped more like a canine tooth than a claw. It is linked to the crural gland in the ankle itself, and releases a secretion of peptides similar to the venom of reptiles such as snakes.[12] This is yet another example of convergent evolution, adapting the same chemistry in the body for similar purposes, but in completely unrelated lineages. Echidnas also have the spur, but they lack the venom.

Venom is not unknown in other living mammals, but it usually comes out of the other end. Venom is a type of poison, but whereas poisonous animals do the deed when you touch, eat or inhale their noxious chemistry, venomous animals inject theirs directly into the skin. Snakes and scorpions are famous examples, but there are also fish and amphibians that employ venom for attack and defence. Among placental mammals the shrews are the guilty party. Some species have venom present in their bite, which may be used to incapacitate insects to store them through the winter.

Perhaps the most intriguing venomous mammal alive today is the *Solenodon*. This creature can grow to nearly half a metre (22in) from nose to tail-tip, and looks like a shrew on steroids. It has a coat ranging from almost black to hay-yellow and even chilli-pepper red, and the hairless tail resembles that of a rat. They only live on the islands of Cuba and Hispaniola, and represent one of the oldest branches of the modern placental mammals still alive today, having diverged from the rest around 70 million years ago in the Cretaceous. Exactly when they diverged, and how they reached the only two islands they now call home, is still a matter of debate. They may have been carried there across the sea on a raft of vegetation, or reached it when the islands were connected to the mainland. In their long pointed snouts, *Solenodon* have two grooved lower incisor teeth. It is along these grooves that their venomous saliva is delivered.

Naturally the platypus just has to do things differently. Like a frisky evil cowboy, the male platypus secretes venom from its ankle spurs during the mating season. This suggests they are used in competition with rivals rather than defence, but humans who have been unfortunate enough to experience their sting report it to be excruciatingly painful – apparently it can even kill a dog.

Intriguingly there appear to be several venomous mammals in the fossil record. Two mammals from the Early Cretaceous, *Gobiconodon* and *Zhangeotherium*, possess a spur on the ankle. Evaluating the Mesozoic mammal fossil record as a whole, Jørn Hurum and co-authors Zofia Kielan-Jaworowska and Zhe-Xi Luo (we will come across the latter two key figures in mammal research again later) found evidence for structures that supported similar spurs in several more species.[13] They didn't find them in the non-mammalian cynodonts, however.

The fact that living monotremes and extinct Cretaceous mammals share the trait of having a venomous structure in the hind foot suggests it may have been a feature of their common ancestor in the Late Triassic or Early Jurassic. Indeed, it may have been widespread in all of the earliest mammals. Only later did placentals and marsupials lose their spurs, perhaps when they grew large enough to defend themselves by tooth and claw instead.

If they manage to survive being hunted, we think of small mammals as living fast and dying young. The common shrew for example, *Sorex araneus*, lives for less than 16 months, and even the most dearly loved pet rat will break your heart when it is laid to rest after three or four short years. But recent research suggests this wasn't the case for the very first mammals.

There are rings in our teeth formed from something called cementum. It mineralises in patterns of growth and cessation annually, just like rings in a tree trunk. Cementum rings have been used by archaeologists to study ancient humans, and by zoologists to determine age in wild animal populations. Using synchrotron X-ray scanning – a technique transforming palaeontology, as we will discover in the next chapter – a team of palaeontologists led by Elis Newham found that the earliest mammals from the Late Triassic and Jurassic lived for over a decade,[14] veritable ancients by small mammal standards. Little *Morganucodon* reached as much as 14 years old, an unachievable goal for most modern small mammals. It was only in the first members of more modern lineages that lifespan seems to shorten, perhaps linked to a rising metabolism.

Although at first glance they may have looked like typical shrews and rats alive today, the Late Triassic mammals were clearly quite different, biologically and anatomically. It wasn't just their spurs that made them cowboys; there was a touch of the John Wayne in the way they walked too. They still hadn't fully brought their limbs underneath their bodies, as in all modern descendants.[*] They probably still laid eggs, and although they produced milk it was unlikely to have been delivered via a teat (more on this later). Their long lifespan suggests their physiology was unlike that of mammals alive today, possibly growing more slowly and maintaining a slightly lower body temperature.

Although the assembly of recognisable modern mammal features was well underway, most aspects of their biology were still being tweaked. They were mammals, but not as we know them. Not yet.

The first mammals of the Late Triassic and Early Jurassic were pioneers. They did something no dinosaur could do: they shrank and exploited an untapped night-time niche. To this day, of the 5,500 or so mammal species alive on Earth, 90 per cent are small-bodied, most of them rodents. The median body mass for mammals today is less than 1kg (2.2lb).[15] Being small scuttlers wasn't an evolutionary relegation, but

---

[*] Except for some exceptions like monotremes, but their strange posture might have evolved from a more upright ancestor.

such a great way to make a living that they are still doing it over 220 million years later.

Putting their newfound bite to good use, mammals became the scourges of the insect world. Hunting in the relative safety of night with senses enhanced, the first mammals grew bigger brains, paving the way for increasingly complex social interaction and behaviour.

These tiny ancestors were living microchips. They were night-vision goggles. They were fuzzy little ninjas, wielding shuriken teeth to reap their insect prey in silence and stealth.

Beware the predator foolish enough to mess with these small warriors.

CHAPTER EIGHT

Digital Bones

Not so much rock as a stone womb,
preserving life lived, waiting for discovery,
pick and chisel replaced with discerning scan,
today's technology birthing
this 3D re-creation ...

Fiona Ritchie Walker, *After Life: Finding Tiny*

The hillwalker pushes those last few painful steps to the top of the mountain, calf muscles burning with lactic acid. The path peters out onto a summit tonsured by a thousand Vibram soles. Her husband catches up with her, his titanium walking poles clacking. They pause to drink in the deep, U-shaped valley beneath them like an open palm with a town sitting in its centre.

Grenoble, at the foot of the French Alps, sits in a valley fringed with pine forest. A river meanders along the flat basin and a road snakes beside and over it, into the town which spreads southwards as though spilled from a cup. Caught in the river's fork is a massive steel circle. From the mountaintops it looks like a polo mint held between a child's fingers. This building dwarfs the structures around it, and its white roof gleams in the sunshine.

You might guess it was a sports stadium or colosseum, nestled in the ice-carved valleys between contorted mountains. But this gargantuan wedding band is the European Synchrotron Radiation Facility (ESRF).

You know you hear those scare stories about the facility in Switzerland that physicists have built underground and are using to whirl particles around, inevitably creating a black hole or ripping into another dimension or ending time itself and destroying the universe? ESRF is a bit like one of those. The difference is that while the one in Switzerland (CERN, the Conseil Européen pour la Recherche Nucléaire, or European Organisation for Nuclear Research) whirls and smashes protons together, the synchrotron whirls electrons instead. It uses them for everything from understanding the structure of enzymes, to finding out why refrozen ice cream tastes rubbish.* It is also providing unprecedented new data on fossil organisms, transforming the practice of palaeontology completely, and exposing our mammal origins.

The main building at the ESRF looks like a cubic crystal. The utilitarian façade opens into a glassy, four-storey atrium. I was led there in the spring of 2017 by Vincent Fernandez, who was giving me a tour of the facility during down-time between scans of my fossil mammals from the Isle of Skye. Fernandez, a French palaeontologist only just in his thirties, has already made a name for himself internationally as one of the world's leading experts in using synchrotron X-rays to produce exquisite scans of fossil material. He worked as part of a team led by Paul Tafforeau, who pioneered the use of synchrotron radiation for palaeontology. Tafforeau first used the technique to examine the teeth of fossil primates, and in the course of his research he applied the same techniques to many other groups. Now he is one of the most important scientists at ESRF, and his influence has resulted in a research stream at the facility focused almost entirely on palaeontology.

Fernandez was one of Tafforeau's team when I first met him. He is famous among palaeontologists for his skills, particularly the digital visualisations of the data he collects. One of his most iconic pieces of work is the reconstruction of a unique specimen of the cynodont, *Thrinaxodon*. This particular animal was entombed inside its burrow when it flooded. From the outside the burrow is nothing more than a tubular lump of russet rock, dug from the Triassic sediments of South Africa. Thanks to the penetrating gaze of the synchrotron, Fernandez could digitally reconstruct its skeleton still inside the stone, without removing a grain.[1]

His scans revealed a creature curled up, its spine curved in an S-shape, with feet tucked underneath a long body. The head was

---

* It turns out that changes in temperature alter the ice cream's microstructure, forming large crystals and coarsening the texture, which impacts on our perception of flavour. Whatever else is happening in the world, it's good to know this important global issue is being addressed.

resting to one side, like a cat snoozing in front of a fireplace. This image, surprising enough in its haunting beauty, was more surprising because *Thrinaxodon* didn't die alone. Sharing the burrow was an unlikely bedfellow: *Broomistega*, an ancient amphibian named after palaeontologist Robert Broom. This *Broomistega* had several broken ribs, which may have driven it to take shelter in the burrow. It is thought that the *Thrinaxodon* was in a kind of hibernating state called aestivation, perhaps toughing out the hot season. Whatever drove them to their burrow-sharing, when a flash flood swept through, neither animal escaped.

In the airy atrium, Fernandez led me to a table-top scale model of the synchrotron. He explained in his soft vowels how it worked.

Like Senator Ted Stevens' description of the internet,* the synchrotron is a series of tubes. It starts with an electron gun called the linear accelerator, or linac – a starter pistol that fires at 200 million electron-volts. It shoots into a circular tube called the booster synchrotron, which is 95 metres (311ft) across. The booster is hidden under a grassy knoll, not visible from the outside.

Inside, the initial gunshot of electrons speeds up, reaching six billion electron volts. As Fernandez described it I imagined time-lapse footage of cars on a roundabout at night, their headlights whizzing in circles. These electrons, now sped up, are injected into a larger storage ring encircling the booster. It is the building that houses this storage ring which can be seen from the surrounding mountaintops: a giant white doughnut on the valley floor, 844 metres (2,770ft) in circumference.

Inside the storage ring electrons zoom at nearly the speed of light. They whirl around for hours, occasionally being topped up again from the booster to ensure they stay at a more or less constant energy and speed. Around the storage ring are a series of magnets, and these

---

* On 28 June 2006, Senator Ted Stevens of the United States said: 'The Internet is not something that you just dump something on. It's not a big truck. It's a series of tubes. And if you don't understand, those tubes can be filled and if they are filled, when you put your message in, it gets in line and it's going to be delayed by anyone that puts into that tube enormous amounts of material ...' The real surprise wasn't that a man in his eighties didn't understand the internet, but that he was part of the committee in charge of regulating it.

act on the electrons to produce X-rays. It is these X-rays that can be used to scan fossils, looking through rock and into the bones themselves to illuminate their structure in unprecedented detail.

Entering the storage ring building, it looks like some kind of science warehouse, packed with props, cables, wires, metal struts and pipes. It's brightly lit and kept cool – despite the unusually hot April weather – with endless air conditioning. A walkway suspended above the industrial chaos allows you to look down on the roofs of cabins arrayed around the ring. Each one of these cabins is built to house a beamline: the spokes flung from the Catherine wheel of the storage ring. The X-rays are the sparks, and they are directed down these beamlines and passed through objects placed in their path.

Looking into the beamline cabins is like peering inside the brains of physicists. The contents are all shiny metal, tangled wires and a disarray of buttons and complex devices. Experiments are taking place. To use these beamlines, scientists like myself apply to ESRF outlining what we want to find out, what the most appropriate beamline is (they all differ in strength and capability) and how much time (beamtime) it will take to scan our material. Our applications are considered by a panel, and those judged to be scientifically sound and important are granted time on a beamline. Getting beamtime isn't easy, and if you are successful you're given shifts that run 24 hours a day – and you are expected to stay up all night and use them.

I counted myself lucky to have Fernandez fighting my scanning corner. He was using his expertise to acquire images of my Scottish mammal fossils that no one else had been able to achieve. ESRF likes to grant projects in which conventional scanning can't do the job – and my Skye mammals fitted the bill. Three regular micro-CT scans at institutions across Scotland and England had failed to provide any detail of the tiny bones and teeth in the limestone blocks. The blocks were too large, and regular scanners weren't powerful enough. Without detailed images I wouldn't be able to reconstruct the skeletons, or study their petite anatomy. The synchrotron was my last hope.

The whole ESRF facility had the feeling of a set piece for a movie. I expected to see white-coated scientists swishing absent-mindedly down the hallways, muttering to themselves. Through glass walls were desks stacked with books and papers in haphazard piles, and whiteboards scrawled with formulae. It wouldn't have surprised me if some zombie or mutant plague was let loose from a basement laboratory. Fernandez and I would have to make our escape, back-to-back, through the polished hallways. There was plenty of room in the atrium to swing a nail-studded baseball bat.

The truth was duller, but an obvious relief. The scientists strolling through the building and operating banks of supercomputers varied from biologists to physicists to crystallographers. Some were employees from food companies and production industries; others were blue-skies researchers. Although there were certainly plenty of 30–50-year-old white men, there were people from international backgrounds, and a healthy smattering of women.

I asked Fernandez what would happen to the synchrotron in a zombie apocalypse. He paused for a moment to check if I was serious, then told me that without fresh electrons being injected from the booster ring, the storage ring would probably be empty of electrons in a couple of days. There would be no Chernobyl-esque explosions or gross mutations in the local wildlife. I made a mental note not to bother coming here for an endless source of energy – or three-eyed deer – in the post-apocalyptic future.

The control room at ID19, the beamline Fernandez and I were using at ESRF, looked like a storage space for computer screens. Seven monitors lined the desks along one wall, and at the far end hung a massive flat screen. All of these were connected either to powerful computers, or the beamline equipment itself. This control room was where the beamline was operated and monitored, and the scans it yielded were reconstructed to show fossils in three dimensions.

It wasn't all high-tech. On a shelf above the monitors sat four old televisions the size of suitcases. Paper is not dead either: a small library of notebooks littered the shelves, and folders with cryptic titles like *ELMO* and *ROBOT 2015* written down their spines. On the table-tops, alongside pieces of wire, metal components and various technical paraphernalia, lay scatterings of cardboard and Styrofoam. High-tech equipment like a synchrotron sometimes requires low-tech solutions like sticky tape.

Fernandez and I placed one of my fossil blocks into a cradle made out of the corner of a Styrofoam box, and secured it using ESRF-branded sticky tape. We carried it next door and stuck it in the path of the beamline X-rays – which were turned off while we were in the room – gluing it down on a bed of double-sided sticky tape. Such simple solutions; it was like *Blue Peter** does physics.

Simple is the last word I'd use to describe the inside of the experiment room though, where the beamline emerges after leaving the storage ring and passing 145 metres (475ft) underground into the ID19 building. Not all of the beamlines are housed in the main circular building. ID19 is one of the furthest from it, sitting on the other side of the road in a bunker half-submerged under tarmac and grass.

The experiment room was packed with equipment. Much of it is incomprehensively mechanical and technical to a non-engineer like myself. Every wall had workbenches bedecked with tools. Coming out of the ceiling were pipes that ended halfway down the wall in red-handled taps. They had alarming labels like *Nitrogen* and *Helium*. Others only had numbers. There were warnings: *Radioactive. DANGER! Avoid eye or skin contact.* Every surface was coated with buttons, some of them friendly green ones, or small yellow things with puzzling words like *Bypass* and *Laser PS*. Then there were the giant, palm-sized buttons with EMERGENCY STOP emblazoned on the top. They both alarmed and reassured me: I hoped never to press one, but if I did there was no chance of missing the damned thing.

Once the sample – the Skye fossil – was in place, Fernandez and I left the experiment room, swinging the lead door closed behind us like a secret vault. He pressed buttons in a panel on the wall, and an alarm sounded for 10 seconds before the metal door clicked. It was sealed shut.

The room with the beamline is not a place you want to be during an experiment, the radiation would be lethal. Fernandez tells me a colleague once left his laptop in there by accident, and it was ruined in seconds by the X-rays. All visitors to the synchrotron must wear radiation-detection badges, and undergo safety training before they

---

* A children's television series in the UK particularly known for encouraging kids to make things from household items like washing-up bottles.

can use the beamline. It includes not just instructions on what buttons to press to safely enter and exit the experiment rooms, but also sections on radiation risk and cancer-causing and mutagenic agents.

I sat with Fernandez at the bank of computer monitors and watched him type code. An image appeared on-screen and I gasped. A vertebra from my Skye mammal was clearly visible in the rock. Even the structure of the bone could be discerned: it looked like it was made out of honeycomb. It was better than I could possibly have hoped.

The X-rays hurled down the beamline to ID19 are wigglers. This refers to a process where a series of alternating magnets cause the electrons to jiggle from side to side as they pass by. It releases much more intense X-ray radiation than is produced by simpler bending magnets, and it can be focused for the specific needs of beamline users. The X-rays shoot out of the beamline and hit the specimen – in this case a fossil vertebra – interacting with the material. A detector panel behind the specimen picks up the projection caused by the X-rays.

Think of hand shadow-puppets in front of a lamp. You place your hand in front of the light and cast a rabbit shadow on the wall. The wall is your detector panel, where the pattern of shadow is collected by the computer. Because X-rays pass through materials in different ways depending on what they're made of, the shadow that hits the wall is altered. The genius of the technology and those who wield it with skill, is first in interpreting the alternating shadows to see inside the specimen, and second in rotating the object to get a series of these shadows. They are then knitted together, digitally reconstructing the whole object in three dimensions, inside and out.

Fernandez is one such skilled user. His delicate fingers flit across the keys as he calculates the path of the beam through my fossil, and how the projections will fit together. In the end it will take 10 hours to scan the most important part of the fossil at six microns in resolution. One micron, or micrometre, is 1×10^{-6} metres, or one-thousandth of a millimetre (0.000001 metres). In other words, the resolution of my scan is the size of a bacterium.

That's a damn good scan. When we're done, I can finally start bringing the Jurassic mammals of Skye back to life.

The mammal fossils of Skye are a unique part of Scotland's natural heritage. Like all fossils collected on the island, they must be deposited in a museum for the benefit of the nation and future scientific study. This is not only a stipulation of our permits to collect them, but the ethical thing to do. I must admit, though, that I feel especially connected to these fossils because they have been the focus of my life for several years, forming the core of my PhD and ongoing research into mammal origins.

The mammals from Skye were the first Mesozoic mammal fossils found in Scotland. They were discovered in 1971 by a palaeontologist-turned-schoolteacher named Michael Waldman. A brief paper about them was published a year later, with his collaborator, Robert Savage. Apart from that, the specimens were virtually unstudied until now.

Amazingly, two skeletons were among the untouched material collected from Skye – incredibly rare discoveries, and even more so at that time. Part of the reason for this lack of attention was that these fossils had been mislaid. It can happen easily, especially when specimens are moved around or curators retire. After several years during which they were partially prepared, moved around, and then lost, it was a decade before they were relocated. For the next 10 years my supervisors at National Museums Scotland had looked for someone to study them. That lucky person ended up being me.

There was much mystery over the details of the fossils' discovery. All of the material found on Skye was supposed to come to the National Museum of Scotland (then called the Royal Scottish Museum), but some specimens had accidentally been retained at the University of Bristol, where Savage had been teaching and researching for most of his life. Where exactly had the two skeletons come from on the Skye coastline, and how did they get back to the museum in Scotland?

As well as studying them, I decided to try and retrace their history, following their journey from Hebridean coastline to Edinburgh collections centre.

I began following the breadcrumbs of Waldman and Savage's path to the Isle of Skye and back again. The trail started in the basement of the Wills Memorial building, the stunning neo-gothic home of the geology (and formerly, palaeontology) departments of the University

of Bristol. Completed in 1925, the church-like edifice is the third-
tallest building in the city, looming at the top of calf-bustingly steep
Park Street. The Wills has become a symbol of the university, its
Great Hall hosting the graduation ceremonies of countless cohorts.
The main tower houses Great George, a 9.5- tonne bell with its own
Twitter account.*

Most of the palaeontologists have now relocated to a swanky new
building nearby where they can rub academic shoulders with biologists
and zoologists rather than volcanologists. But at the time I studied my
masters there, the Wills Memorial remained the seat of palaeobiology†
teaching. It seems every second palaeontologist in the United Kingdom
has studied, taught or worked at Bristol at some point. Spearheaded by
some of the most prolific names in palaeontology, it attracts a studentship
from around the world to study all aspects of extinct life on Earth.

Inside, the curator of the university geology collections led me on
my quest. I walked past the theatre I'd spent so many hours in, where
the spicy smell of warm wood would fill the air during hot summer
lectures. We descended into the bowels of the building. Boxes of
rocks were stacked against the walls. We wove between them, turning
corners. There was a deep humming like a Buddhist chant – the
churning of the heating system and air-conditioning that served as the
century-old structure's lungs. Via a warren of corridors we came to a
room stacked with collections. Fossils, rocks, books, cabinets; a jumble
sale of natural science that stretched to a high ceiling.

The curator led me to the back of the room where he began
shuffling stacks of large plastic storage containers. On top of a wooden
cabinet beside us grinned the massive skull of *Megistotherium*. This
carnivorous creodont lived over 15 million years ago and had crushing
teeth that would have delivered a powerful bite. With a skull over half
a metre (20in) in length, it is one of the largest of a family called
Hyaenodontidae that were among the dominant carnivores before
modern Carnivora (cats, dogs, bears and the like) evolved and took

---

* twitter.com/greatgeorgewmb
† Palaeobiology is more or less synonymous with palaeontology, although
arguably with a heavier biological component. This reflects the increasingly
interdisciplinary approach taken by palaeontologists.

over that niche. Savage had named *Megistotherium* in 1973 from specimens he collected during fieldwork in Africa – the research for which he is best known.

Five boxes in the university stores held the notebooks, papers and correspondence of R. J. G. Savage. Some of it was left in his office and never claimed, and the rest had been given to the University for safe-keeping after his death from cancer in 1998. I was left alone with the boxes at a small wooden desk in the corner.

Opening the clip lids released the comforting musk of ageing paper. Several beetle carcasses littered the box – a museum curator's nightmare, as they belonged to creatures hell-bent on eating history. I brushed them off and shuffled through typewritten manuscripts, peppered with scrawling handwriting. Most of it was about the African fossils he specialised in. There were high-quality photographs of bones, drawings and diagrams – the initial stages of what would later be figures in published papers.

A newspaper clipping captured the young Savage in 1953, crouching down to inspect the skull of an 'Irish elk', *Megaloceros giganteus*. They became extinct about 8,000 years ago. This particular specimen was found in the peat of a small loch in Northern Ireland called Lough Beg. Savage was asked to examine it because he was one of the country's foremost mammal palaeontologists. He might have informed the local men who found it that this so-called elk was misnamed: it was not closely related to either the elk or moose (*Cervus canadensis* and *Alces alces*), but was a completely separate species of extinct deer.

In the 1950s, many scientists still believed the giant antlers of *Megaloceros* – the largest ever recorded on a deer, at up to 3.5 metres (over 11ft) from tip to tip – were the cause of this animal's demise. Surely, no creature could survive with those weights beating them down? We now realise this was not the case. Like other deer species, the male *Megaloceros* had undoubtedly grown their head adornments to compete for mates, for sexual selection. Why they died out is still unclear, but ecological changes, disease or hunting by humans are far more likely.

It takes three men to hold the enormous skull. Savage, in a stylish wide-trousered suit, his hair slicked back and round glasses perched halfway down his sharp nose, cradles the underside of the skull and peers at it, as though reading its long-gone mind.

Alongside the papers, several small wooden boxes opened to reveal unexpected zoological treasures. One contained thin-sections of mammal teeth: sheep, walrus, hippopotamus, juvenile platypus.* Another was Savage's collection of mammal middle ear bones, all that remain of the multiple bones that once comprised our ancient ancestors' jaws: malleus, incus, stapes.

Finally I found what I was looking for: several tattered hardbound notebooks. The edges were bashed and the binding loose. On the spines Savage had written his field locations in black marker pen: 'Kenya', 'Tanzania', 'Libya'. I searched to find the one I sought.

'Skye'.

The notebook was once red, but now faded to pink and tattered at the seams. 'Skye R. J. G. S.' it read on the front cover. It smelt like an abandoned house, or an old boot. It may have only been 40 years old, but it felt like an ancient relic. Opening the pages I found pencil squiggles that were hard to interpret, the hieroglyphs of another generation. I struggled to decipher his hasty symbols.

Finally I found the entry I'd waited so long to find.

> Monday 11th of September, 1972: ... found associated skeleton of mammal in lone block on beach below landslip. Proof this section needs thorough searching.

Savage, Waldman and their team had found the first near-complete Mesozoic mammal skeleton in the British Isles, on Skye. At the time, it was one of the most complete Jurassic mammal fossils in the world.

Waldman was a vertebrate palaeontologist, having studied geology and zoology at the University of Bristol under Robert 'Bob' Savage – working on the iconic *Megalosaurus*. For a year, Waldman worked as a diamond prospector in Liberia, before carrying out graduate and postgraduate research in the 1960s in the Canadian Badlands and in Australia, studying dinosaurs and fish. After returning to Great Britain he became a geology teacher at Stowe School in Buckinghamshire, continuing his research alongside teaching until his retirement in 2002.

---

* They have the remnants of teeth when young, but lose them later.

Affectionately known as 'Doc Pot' by his pupils, Waldman was known for his enormous rock collection, which he had donated for teaching geology. He inspired the pupils to appreciate fossils, stones and minerals, encouraging those who were interested to pursue geology – often at the University of Bristol under Savage, once his teacher, now also his friend. Waldman was clearly well liked by the pupils at Stowe, some of whom sent him postcards years after they'd left, depicting rock formations or volcanoes they'd learned about in his classes.

In 1971 Waldman and a fellow teacher were due to visit a school group doing their Duke of Edinburgh Award, camping at Camasunary in Scotland – a well-known beauty spot at the foot of the Cuillin mountain range on Skye. He decided to read up on the area beforehand, hoping for fossil-bearing rocks. He found several references to fragments of bone on the Strathaird Peninsula, recorded in the early 1900s by a team from the Geological Survey. Waldman knew from previous experience that where you found a few vertebrate fossil fragments, you were likely eventually to find more. It was time someone took a second look. Fortunately the path to the camp would allow for a small detour to investigate.

Waldman was right to check again. When they reached the shore they saw tiny fossil fragments peppering the outcrops of Jurassic limestone. For half an hour they found increasingly interesting fossils. Finally they discovered the first complete limb bone. It was the length of a matchstick, and belonged to something quite small – a little crocodile perhaps. Excitement building, they searched on. Finally Waldman discovered something so unlikely, the hairs on the back of his neck stood on end as he stared at it. There were double exclamation marks in his diary entry that night: 'This appears to be mammalian … I hope it really is a mammal jaw!!'

No one had ever found a Mesozoic-aged mammal in Scotland before. Could it really be?

The north-west of Scotland and the Western Islands are famed for their achingly ancient tectonic thrusts, jagged peaks of volcanic burps, and landscapes scraped and shaped by glaciers. None of these were prime vertebrate fossil spots. No one paid much heed to sedimentary rocks in those regions. Apart from a few marine reptiles found on the

Isle of Eigg by nineteenth-century Highland palaeontologist and writer, Hugh Miller,[2] the Hebrides were not really known for vertebrate fossils.

Fossil mammals from the Mesozoic, globally speaking, were still desperately rare. By the 1970s isolated teeth were known from a handful of places around the world, and fragments of jaw or skeletons, like *Megazostrodon* from South Africa, were among the rarest of all. It was said at the time that all the fossil mammals collected around the world would barely fill a single shoe box.

Two days after their initial exploration of the site, Waldman left the pupils in the care of his colleague and hiked back to take another look. He examined his prize find – the enigmatic jaw – under the hand lens. He argued with himself in his diary, setting down the reasons it could belong to a shark or a reptile instead. But those complex cusped teeth, their double roots in that straight fragment of jaw bone ... He knew enough about reptiles and fish to know they didn't fit the bill. Dare he hope?

With utmost care, Waldman collected the little fossil from its pale grey bed on the cold Skye shoreline, and hiked back to camp.

Returning south, Waldman urgently contacted his former Professor at Bristol, Robert Savage. By all accounts, Savage was an unusual man, perhaps one of the last of the 'gentleman naturalists' who once dominated the natural sciences. Born in Belfast in 1927 to a moderately wealthy family, he studied geology and zoology at Queen's University before moving to London to complete his PhD describing an extinct otter, called *Potamotherium*. He ended up at the University of Bristol in 1954, where he remained for the next 32 years: first as a lecturer in geology and curator of the geology museum, later as professor of vertebrate palaeontology.

Under Savage's influence, the University of Bristol's palaeontology department thrived. In 1969 Savage founded the joint honours degree in geology and zoology at Bristol. This was the foundation stone of what is currently one of the largest palaeobiology research groups in Europe, welcoming an international cohort of students.

When I ask acquaintances what he was like, they always use the same word: *gentlemanly*. Savage often invited his students to his house for meals. It was a beautiful home, filled with curiosities of geology and palaeontology. A 'gentleman's bachelor pad'* in the wealthy area of Clifton, as a visitor remembers it, with antiques spread across several floors. Savage welcomed guests, often hosting researchers from across the country. He would give them a whole floor of the house for their own use, and invite them to help themselves to his rich supply of brandy and Scotch.[3] The palaeoanthropologist Richard E. Leakey, Savage's friend and colleague, recalled how Savage would present visitors with obscure fragments of bone and see what they made of them. He particularly loved occasions where they guessed correctly. His students were also expected to play this game.

On some occasions the guessing extended from fossils to food. One of Savage's former students told me about a meal he provided for her cohort.[4] They arrived to find a hearty stew waiting on the table. As they ate, Savage asked them what they thought it was. Venison perhaps? It seemed quite meaty. After much guessing, their professor revealed the main ingredient was baby hippo. Through his ties with keepers at the Bristol Zoo, Savage had apparently been given the animal's body (it died of natural causes) so that he could keep the bones to add to the teaching collection in the University of Bristol. As it was fresh, he'd decided not to let so much good meat go to waste.

In 1971 Waldman brought the tiny fossil jaw from Skye to Savage in his home in Clifton, and handed it over. Robert Savage looked at it for some time with the hand lens. Suddenly he leapt from his chair and began pacing around the room. 'Do you know what you've found, Mike?' he exclaimed, 'It's a mammal!'

A week later the pair of them bundled themselves into a Land Rover and pointed the bonnet northwards.

Skye was a great deal harder to reach in 1971. They arrived in Kyle of Lochalsh from Bristol at the end of July and took the small car ferry over the short stretch of sea. On the other side, Waldman stopped the local postman to ask for directions to their accommodation in the village: 'We're looking for Mrs MacKinnon'.

---

* Savage wasn't always a bachelor, he married in 1969, but his wife Shirley Cameron Coryndon – who was also a palaeontologist, specialising in fossil hippopotami – died in 1976.

The postman regarded him sympathetically. 'Which Mrs MacKinnon?'

'Mrs Mary MacKinnon.'

The postman sighed, 'Well, that doesn't really narrow it down I'm afraid.'

'She has a husband named Angus,' Waldman elaborated.

'Ah,' the postman replied, 'well that narrows it down to three then.'[5]

Eventually locating the correct MacKinnon, Waldman and Savage had a restless night. They woke early and gulped down breakfast as fast as possible, then left the Land Rover at the post office and trekked to the site where Waldman had found the mammal jaw.

They were rewarded immediately. Savage stated in his sparse geologist's field diary, 'Mammals: teeth and jaws, when fresh, well preserved. Often bone is so weathered, unidentifiable.' They may have been weathered, but that didn't dampen the pair's enthusiasm. They spent the day scouring the shore, amazed by the wealth of bones that had lain undiscovered for so long. There were turtle shells, crocodile and shark teeth, but more wonderfully, the scattered jaws and limb bones of small Jurassic mammals and lizards.

The following April, Waldman and Savage returned with an 11-person team from universities across the UK. They spent two weeks searching sites across the Isle of Skye from the northern tip to the southern inlets, hunting for fossils and making detailed notes on the geology. In the second week they visited the Isles of Eigg and Muck, seeking Hugh Miller's 'Reptile Beds'. In September of the same year, Savage, Waldman and two others returned again to Scotland, visiting what was then called the Royal Scottish Museum in Edinburgh on 6 September. They examined 12 drawers of Hugh Miller's material, before travelling north to Skye again.

This time, they really struck mammal gold. They found not one, but two mammal skeletons, and a wealth of other material including the remains of fish, crocodiles and turtles. They had uncovered one of the most important Middle Jurassic fossil vertebrate localities in the British Isles.

However, the significance of their discoveries wouldn't be fully realised until much later. Impressive enough at the surface, the fossils of Skye would go under-studied and under-appreciated for the next three decades. Only then did scientists like me and my colleagues have the technology to look beneath their cold stone surface to reveal their long-held secrets.

Using CT scanning, computed tomography, is a ubiquitous part of modern palaeontology. This is especially true for vertebrate animals, but it can be used for the study of invertebrates, plants and rocks themselves. For most of us our first and perhaps only encounter with CT scanning is medical – after an accident for example. We know about the process from television medical dramas, where the patient slides into a circular hole in a room-sized machine. A team of doctors in an anteroom point at their insides, lit up on screens, and diagnose them. Artistic licence aside, this is what computed tomography was developed for. It is a type of X-ray machine, making it possible to see the internal structure of objects, including people.

The CT scan, also called a CAT scan (computerized axial tomography), was developed in the 1960s, and the first commercial medical scan was carried out in 1971. The two people who were key to developing it – a South African-American physicist named Allan MacLeod Cormack and an English electrical engineer called Godfrey Newbold Hounsfield – won the 1979 Nobel Prize in Physiology and Medicine for their efforts.

Building on the developments of the preceding decades, the CT scan worked on the principle of radiodensity. This is the understanding that portions of the electromagnetic spectrum pass more easily through some materials than others. Materials that don't allow electromagnetic radiation to pass through them are radiodense, those that do are radiolucent. This transparency is measured in Hounsfield units (HU). Distilled water measures 0 HU, air measures -1000 HU, and bone varies from +400 to +1000 HU. The trick for commercial application was to find a way to use this knowledge of radiodensity to see through an object three-dimensionally.

X-rays are a natural part of the electromagnetic spectrum. At one end of the spectrum the longest wavelengths include visible light – the world we see around us naturally with our (somewhat limited) mammal eyes. Longer still are infrared, microwave and radio wavelengths. At shorter wavelengths are ultraviolet or UV rays, visible to some insects and birds. UV is blasted out from the sun in levels that would be lethal to life on Earth, were it not for the filtering effect of our atmosphere. Even so it is the sun's UVs that cause sunburn and can lead to skin cancer. The X-ray wavelengths come next, used in airport scanners, hospital machines, industry and science. At even shorter wavelengths are gamma rays, which have the shortest length and therefore the highest energy of them all.

In a conventional X-ray image of a broken bone you see the bone as a flat, two-dimensional projection. It is extremely useful of course, but can only tell you so much. The CT scan achieves a three-dimensional image by capturing multiple individual projections, like photographs from different angles, then applying complex mathematics to reconstruct the whole thing. The first scan in 1971 took 180 individual images, each taking five minutes, and the reconstruction afterwards took two and a half hours. These days most of the CT scans palaeontologists acquire of small fossils take less than an hour in total, sometimes only minutes, although it varies depending on the size and nature of the material being scanned.

The results of this scanning and reconstruction process are a series of slices through the object in the x, y and z axes. In other words, we can look at slices from anywhere through an object, be it fossil or person, not only in the horizontal and vertical plane, but also depth-wise. By tweaking the scanner settings we can capture the different radiodensity of the material being scanned, making it possible to differentiate between different types of material: living tissues, air, liquids, rocks and minerals – the possibilities are vast.

Computed tomography has been used in palaeontology since the 1980s. The two earliest scientific papers using it were both looking at mammal skulls: a hoofed mammal from the Miocene (between 5 and 23 million years ago), and the skull of *Homo erectus*. The blurry images in these two publications from 1984 look more like Rorschach tests than slices through fossils.[6,7] But this was a vast improvement on the methods of the previous 80 years.

The biggest advantage CT provided was a chance to see the structure of fossils without destroying them. Manual thin-sections are a long-established part of geology. Invented in the early 1800s for looking at the structure of rocks, they could be (and still are, particularly in palaeohistology) used to study the internal structure of fossils. In 1903 one scientist took the idea a step further.

William Johnson Sollas was the professor of geology and palaeontology at the Oxford University Museum of Natural History in the early twentieth century, and was dubbed 'one of the last true geological polymaths'.[8] He invented the process of serial sectioning, the analogue version of the CT scan. Along with his colleague Reverend F. Jervis-Smith, Sollas created an apparatus that could be used to manually grind away at the surface of a fossil, stopping at set intervals.

'As each section is prepared it is drawn with the aid of a camera lucida, or photographed under a microscope,' explained Sollas in his paper to the Royal Society. 'Photography has many advantages over drawing, particularly as affording a record which may be trusted in questions of dispute ...'[9] Such disputes, should they occur, would be hard to settle once the process of grinding the fossil to dust was completed – 'the fossil is necessarily destroyed in the process', noted Sollas.

Sollas enthusiastically set about grinding his way through every fossil he could get his hands on. Naturally, curators were not keen to have their precious collections undergo the process. However, from the images generated, researchers could create wax models that charted the internal structure of long-dead organisms in unprecedented detail.

As well as destructive, the process was also very slow, especially for large fossils. One palaeontologist famously spent 25 years carefully grinding and recording just one Devonian fossil fish. Serial sectioning was used by Zofia Kielan-Jaworowska, perhaps the greatest mammal palaeontologist of all time, to explore the anatomy of Cretaceous mammal fossils from Mongolia (I'll tell you more about her in the following chapters).

Being able to capture the internal structure of fossils without pulverising them was clearly a huge step forward. Since the 1980s, the images acquired by CT have improved enormously, and the technique has become more accessible. Originally palaeontologists had to find industrial or medical collaborators who would allow them to use their machines. Now, universities and museums often own their own scanner. Rather than filling whole rooms, these machines are more or less desktop: about the size of a household stove.

The type of scans mostly used for fossils are called micro-CT, or µCT. They are more powerful than medical machines, focusing on smaller objects but at much higher resolution, with X-rays powerful enough to penetrate denser materials like rock. But they are also more damaging to organic tissue, so unlike in a hospital scan you can't use micro-CT on people without causing tissue damage.

These machines resemble microwaves, with thick outer casings to contain the X-rays and a sliding door on the front with a tiny brown window set into it. Inside, the fossil sits on a turntable, and on one side is the gun-like X-ray generator, while on the other is the detector. This picks up the varying patterns of transmitted X-rays and sends them to a computer to be processed into images.

Synchrotron is micro-CT turned up to 11.* It is the superhero of scanning: stronger, smarter and more intense. It succeeds where conventional CT fails, saving the palaeontological day. Synchrotron provides a more intense beam of X-rays, better able to pass through materials and producing results faster. Materials that are hard to differentiate using conventional micro-CT can be more readily discerned. For example the cementum rings researchers counted to find out the chronological age of the first mammals (see the previous chapter) can only be seen using synchrotron. If micro-CT is a run-of-the-mill point-and-click camera, synchrotron is a high-end digital SLR. The resolution, like the megapixels of a camera, is greater, and this means it allows researchers to investigate much smaller objects in great detail.

Also like photography, you can have the best equipment in the world, but that doesn't mean you automatically produce great results. Using the right exposure time, length of scans, distance from the specimen and many other settings on this complex machinery to suit the material you are scanning, is a skill in itself. Then there is the cost

---

* From the 1984 movie, *This Is Spinal Tap*. When asked if the amplifier dials going to 11 instead of 10 is louder, the band member explains, 'Well, it's one louder, isn't it?'

of the equipment, of scan time (which is often charged hourly or daily), and getting access to the machines themselves.

All of this means that you can't just pop your ancient bones in and wait for the ping. It is not a magic bullet. But when used correctly, it produces magical results.

As I endured the second night of not getting to bed until after 2 a.m., even the novelty of the *DANGER!* buttons wasn't enough to maintain my enthusiasm. I hit the pillow like a lead weight on the sea-bed. The sediment of the day settled as I drifted off.

The Skye mammal fossils are slowly revealing themselves, but the many thousands of projections (individual scans) that accumulated at ID19 during my visit were just code on a server. It would be months before Fernandez would be able to send me the completed scans. First he had to reconstruct the projections we took; then, because my fossils were too big to capture in one go, he would stitch together the multiple scans we'd made. Once they were combined he would create the slices I needed for the next stage of reconstruction: digital segmentation. That in itself would take many more months.

Although you can study fossils using CT slices alone, it is digital segmentation that has really transformed our science. This is the process of taking the slices and putting them into a program to turn them into a three-dimensional object. It is the computer equivalent of Sollas's wax models. When you see amazing images of scanned fossils, it is the segmented digital reconstructions you are looking at, not the original scan data.

The digital reconstruction of fossils is part computer process, part art. It can be heavily automated using in-built software tools. Many of these rely on converting greyscale values in your data into a 3D object. By choosing what greyscale values you want, you can automatically select your fossil from out of the encasing rock. It is almost like using a cookie cutter.

But fossils are rarely so straightforward. How easily you can digitally remove them from the rock depends on how much contrast there is between the fossil and the stone. If the scan wasn't done optimally, the same problem arises. If the greyscale values are too similar, when you try and select the fossil you'll include lots of rock, resulting in a useless blob on the screen. Even in the very best scans it is often hard to differentiate between stone and bone. In these cases it is down to the skill and patience of the researcher to go through the data and figure out what's what. This introduces some subjectivity, but on the whole researchers are cautious people, taking weeks and even months to work their way through, picking out the bones in the digital realm. This was the fate that awaited me on my return to Scotland.

The results of research using scan data can be breath-taking. One study that really showcased this for mammals was carried out by French palaeontologist Julien Benoit and his South African colleagues. In 2016 they published a landmark study[10] using micro-CT to explore structures in the skulls of therapsids, cynodonts and mammals. Their work revealed something we thought could never be known: the origins of whiskers, fur and milk.

Because hair rarely fossilises, to trace its origins researchers must use bones. Some soft tissues leave marks on the bone: muscles are the most obvious example. As we saw in previous chapters, larger muscles have an impact on bone growth, causing it to change shape in response to the pressure exerted upon it.

Other features also have an effect on the body's architecture. Whiskers and hairs are derived from our largest organ, the skin, and they are connected to our central nervous system. Whiskers are mobile in many mammals, twitching as they detect scent or feel their way through the world. To achieve all of this movement and sensation, whiskers and fur require a blood and nervous supply. It is this that researchers used to trace their appearance in mammals.

By scanning fossil skulls from the Late Permian to the Early Jurassic, Benoit's team was able to look at the path of nerves in the skull. The trigeminal and facial nerves are linked to the sensitivity of our faces, particularly in mammals with whiskers. The team traced the pathways and found that at a key point in the fossil record for mammals they were no longer snaking inside the maxilla (the upper jaw), but had emerged through a small foramen (a hole in the bone) and onto the surface of the bone. This meant that their complex branching was taking place just under the skin, which is what you see in modern mammals equipped with whiskers. The fossils in which this first happens are 240 million years old (Middle Triassic), and sit at the base of the cynodont branch that includes mammals and their close relatives.

Nerves don't tell the whole story. A second line of evidence comes from a surprising link between skull structure, hair and the production of milk. In most reptiles and amphibians (and some fish) there is a small hole in the top of the skull called a parietal foramen. This feature is present in our shared tetrapod ancestors, and was retained in synapsids far into their evolutionary history. Associated with the pineal gland in the brain – known as the 'third eye' – it is thought to help regulate animal activity cycles by detecting light levels, and also to help animals thermoregulate.

Apart from the odd few exceptions, the parietal foramen didn't disappear in the mammal lineage until about the same point that Benoit pinpointed the appearance of whiskers.

The lack of a pineal foramen in mammals is linked to a gene that works during development of the embryo. Studies in embryology have shown that when mutations occur in this gene, called Msx2, the results are deeply revealing. These mutants not only have holes in their skull at exactly the same place as the parietal foramen of their ancestors, but in Msx2 mutant mice it causes problems in hair follicle maintenance and disruption to the development of mammary glands. The fact that these three features, pineal foramen, hair follicles and mammary glands are connected by one gene is tantalising. It suggests that in the Middle Triassic, it may have been a mutation in the Msx2 gene in cynodonts that played a role in the development of these defining features of mammals.

Milk itself is a mysterious substance. No other animal group creates a nutrient from their skin in the same way. Some birds, like pigeons, penguins and flamingos, create 'crop-milk' from the cells in their oesophagus, but it is not as rich and copious, neither is it produced for sustained periods of time. There are amphibians that nourish their young through the jelly-like substance around the eggs, or by their young consuming their sloughed-off skin and other secretions. More distantly across the tree of life, there are fish that provide mucous for their young (a detail unsurprisingly not included in *Finding Nemo*), and a type of cockroach that feeds its live-born young on 'milk' from the brood sac.* But these exceptions are minimal in nutritive value or volume by comparison to our mammal wonder food.

Mammal milk varies wildly in composition. Human breast milk is only up to 5 per cent fat, and cows' milk as little as 1 per cent fat, whereas harp seal milk is over 60 per cent fat, plumping up their babies for a life beneath the ice. Blue whales feed their young the equivalent of around three whole bathtubs of milk per day, whereas a dairy cow could only manage a sitz bath (about a third of a tub). The advantage of producing milk is that the young can feed even when the adult diet comprises food that would be too difficult for them to capture or digest themselves, or when the adult is using food stores (including their own fat reserves). As a result mammals can rear offspring in resource-poor environments and through harsh seasons. As long as the adults can feed, their young can survive.

In diagrams, human breasts look like volcanoes.† Surrounded by a bubbly fat layer beneath the sloping skin, the bulk of the boob is made of lobules. That's where the milk is produced. Lobules resemble bunches of purple grapes, fed by a deep network of blood vessels from the body. Each lobule is connected to the nipple by tubes – the milk ducts – and it is of course through the summit of the nipple that milk is exuded.

Researchers have argued about whether mammary glands evolved from apocrine (sweat) glands or sebaceous glands. There is a similarity between them in their basic structures. The sebaceous glands are situated inside hair follicles and coat the hair with oily sebum to keep it healthy. In marsupials, milk ducts often contain hairs which are then lost during development, suggesting an association with hair in the development of milk ducts. Many researchers now suggest

---

* Not only did researchers discover that the Pacific beetle cockroach, *Diploptera punctata*, feeds its young on this pale yellow liquid, but they went on to extract some of it, a process they described as 'milking a cockroach'. Turns out it's incredibly nutritious, so who knows when this delicious new treat will hit the supermarket shelves …

† The human production of milk is one of the things that drove writer Liam Drew to understand what it means to be a mammal in his excellent book, *I, Mammal*, which I highly recommend reading for more about our mammaly biology.

that sweat and sebaceous glands are both connected to the development of mammary glands, in association. Milk contains substances that can be traced back to key proteins for tissue development and immune response, as well as moisturising and maintaining the skin.

It's hard to imagine how something like suckling could develop in the skin from a state of milklessness. There have been over a dozen hypotheses to explain it since Darwin's day. We know that the skin of synapsids was glandular, unlike the scaled skin of reptiles. This is based on the presence of glands in the skin of both mammals and amphibians, which probably both inherited it from their shared ancestor, whereas reptiles lost the feature. Additional evidence is provided by incredibly rare fossils. An exceptional skull of the Middle Permian dinocephalian *Estemmenosuchus* (the one with horns exploding from all over its face that we met in Chapter 5) is rumoured to include one of the few examples of such ancient fossil skin. The research published on this elusive specimen is sparse, but the fossil appears to show that the skin was glandular. This is what we would expect from our predecessors, and provides the starting point for milk production.

The lack of synapsid egg fossils suggests that they didn't have calcified eggs. Even the platypus, our cousin from over 160 million years back, lays leathery eggs with a shell composed of keratin, the substance that makes up our hair and nails. Lizard eggs are also soft and leathery, suggesting it is ancestral for all amniotes. Eggs with soft shells dry out more easily, which is part of the reason that reptiles often bury theirs in moist soil. Some snakes and lizards, such as skinks, have evolved not to lay their eggs at all, but instead hatch them inside the body and then give birth to the live young.

Synapsids were faced with a conundrum: how to keep their eggs moist when their body temperature was increasingly elevated. If they buried their eggs in soil they might stay moist, but would be subject to the temperature of the soil. As the mammal lineage grew more warm-blooded they would have needed to ensure their eggs stayed warm or the embryos would die. However, that same warmth would have increased the chances of them drying out. Echidnas lay their tiny eggs into a pouch to tackle the moisture versus warmth problem, but such small eggs – no bigger than a marble – place restrictions on development. The young are less physically developed when they hatch, and more vulnerable to losing body heat.

The mammalian solution for keeping eggs moist and warm was to sweat all over them. Although I use the word 'sweat', there is an important difference between evaporative cooling using lots of sweat, as humans do, and the secretions of the apocrine and sebaceous glands in other mammals. Most mammals can't sweat to cool down; they cool using other means – for example panting, or licking their skin and letting the saliva evaporate. But they do secrete fluids from their glands, and it is thought that the synapsid lineage began keeping their eggs moist using these skin secretions. Through time, the levels of antimicrobial compounds in the secretions increased, protecting the eggs from desiccation *and* infection. We know this because we can trace the whey compounds in milk back to antimicrobial proteins.

As cynodonts became increasingly small-bodied and laid ever more miniature eggs, their young required more care. (At some point in the Mesozoic at least two lineages – the ancestors of placental and marsupial mammals – stopped laying eggs, instead birthing their young. Exactly when and how this happened remains a mystery). These cynodonts may have begun to feed them on their egg-protecting skin secretions in the first days or weeks of life.[11] Secretions with more nutrients led to healthier offspring, so naturally selecting for richer and richer milk. As we discovered in an earlier chapter, monotremes like the platypus and echidna lack nipples, but secrete milk from a patch on their skin for the young to lap. This is probably how the earliest mammals did it too. The result in later mammal lineages was the wholescale appropriation of the humble sweat gland into a production site for baby food. And thus the nipple was born.

But a nipple is no use if your babies can't suckle. Suckling requires a hard palate, and a muscular throat and tongue. To us it may seem natural that all animals have the same ability to manipulate food in their mouths, but it is an especially mammalian pastime. As our ability to chew food improved with the increased complexity of our teeth, our meals stayed in our mouths for longer. To stir this pot of mastication, the tongue was fine-tuned. The bone that anchors it, the hyoid sitting at the base of the tongue, was modified from a simple band (as seen in *Thrinaxodon*) to a jointed saddle shape.

A recent fossil from China is the earliest mammal known to have a modern mammal-like hyoid. It belongs to a tiny Late Jurassic animal called *Microdocodon*,[12] one of the revolutionary docodontans we will hear about in the next chapter. The presence of a hyoid in this fossil from around 160 million years ago tells us that the ability to thoroughly chew food in the mouth dates back to at least this time. With this structure in place, mammals could give their chow a good going-over before swallowing. These manipulations in the mouth may have been co-opted later for suckling on teats.

Thanks to the power of CT and synchrotron, we now know that while giant reptiles were expanding above ground, cynodonts were grooming their newly hatched offspring deep in their burrows. Before long, they were feeding them milk, giving them a head-start in life. As milk became richer, so mammals could become smaller and live in difficult environments, born at increasingly earlier stages of development and brought up on the teat. Tooth replacement patterns changed in tandem, and so the offspring replaced their dentition just once when they were weaned. This had the 220 million-year consequence of human children around the world proudly pulling aside their gums to show us their gappy smiles, waiting for adult teeth to take their permanent place.

Scanning fossils with X-rays provides a non-destructive way to reveal internal structures, but it can also be the key to studying wholesale anatomy. For the smallest fossils, simple observation can't take in the details of even the largest bones. The first mammals were so minute that their teeth are often only millimetres across, like grains of sand. When their bones are encased in rock, removing them from the stone is a risky business: they can be damaged or lost in the process.

The fossil mammal skeletons Waldman and Savage collected from the Jurassic of Skye are only partly visible on the surface. Some of the surrounding rock has been removed, but the rest of their skeletons lie hidden in limestone, too small to extract. The rock itself is dense and doesn't react well to acid, so it can't be easily removed. As a result of all these things, these mammals can only be properly studied using CT.

But there was one thing I already knew about the Skye mammals before I scanned them. Although most of their impressive skeletons were hidden, one important bit was exposed. Their teeth, smaller than the letters typed on this page, were visible. This told me one thing: what *kind* of mammals they were.

It turned out that Skye was home to some of the most important groups of mammals in the Jurassic. Among them were diminutive members once thought almost inconsequential to the story of mammal evolution. We now know that theirs was one of the most experimental tales in the Mesozoic, exploring niches that wouldn't be filled again for over 100 million years.

To find out more about them and the first thriving ecosystems of the Jurassic, I would have to travel to the new frontier of palaeontological science: China.

CHAPTER NINE
Chinese Revelations

This lost two-thirds, this mammalian prehistory ... must contain the answers to the most fundamental problems of mammalian classification and phylogeny, to which later mammals, taken by themselves, always and inevitably yield only equivocal, misleading, or incomplete clues. Only the Mesozoic mammals can cast direct light on these basic early stages, but they have long been either neglected or clouded by misinterpretation or erroneous observation. The reason for this condition of affairs is, however, not far to seek. The remains of Mesozoic mammals are among the smallest, the rarest, and the most fragile of fossils.

George Gaylord Simpson, *A Catalogue of the Mesozoic Mammalia*

The wide space was filled with light and voices. The tinny rhythm of English and Chinese conversations clattered off the pale shining floor tiles and white walls. Outside, Beijing had thrown off its smoggy blanket, and the sun poured across the city and sloshed into the Institute for Vertebrate Palaeontology and Palaeoanthropology (IVPP).

Inside, 20 palaeontologists from around the world were kept cool by overworked air-conditioning in a corner room of the building. They circled a long wooden table like pilgrims at Mecca. Placed along the length of the table were slabs of rock – some the size of a tombstone, others small enough to fit in the crook of your hand. A couple of the biggest chunks were set into plaster surrounds. On the surface of each rock were the skeletons of Jurassic and Cretaceous mammals.

After they had disembarked from the bus at IVPP, then hiked to the sixth floor, the researchers had run into the room like children released in a playpark. Jostling around each specimen, they bent close to the stone, gazing across the miniature landscape of bones and teeth. Multiple intense discussions were taking place: where had each one come from? What did the anatomy mean? Is that cusp really homologous? There were debates on the controversies of different interpretations, heads being shaken, eyes widened. Exclamations of awe and excitement punctuated the atmosphere. Under it all was the constant beeping and clicking of cameras.

For many of us this was the first time we'd seen these fossils in the flesh, so to speak. We had pored over the scientific papers that described them, and studied the minute details of the teeth and bones in order to carry out our own research on Mesozoic mammals. The specimens were key finds published in the biggest scientific journals – 'game-changers' in the field of Mesozoic mammal palaeontology. But it is difficult to access them to study first-hand. The politics of science can be tricky, and for researchers outside China in particular, getting a chance to even glance at these gems is almost impossible unless you are lucky enough to be involved with the research groups at the institutions that curate them.

We were at IVPP as part of the first International Symposium on Mesozoic Mammals. This unprecedented event was organised by members of the Beijing Museum of Natural History (BMNH), IVPP and multiple museums and research institutions across China. The symposium marked the opening of the world's first dedicated exhibition of Chinese mammal fossils from the Mesozoic, held at the BMNH.

As some of the researchers at the forefront of our field, we had been invited to view the specimens before the exhibition opened to the public. Most of these skeletons had only been seen by the curators and research teams who described them. They comprised a roll-call of world-changing discoveries that had revolutionised our understanding of what our mammal kin were up to in the so-called 'time of dinosaurs'. It was also an opportunity to take stock of China's role as a leader of palaeontological science.

The previous day we had been in the laboratory at the BMNH, gazing lovingly at Mesozoic mammals down microscopes and making copious notes. Today it was our turn to do likewise at IVPP. Opening their doors and co-ordinating in this way was an unusual act of cooperation between these major institutions. Competition is fierce, and as in the old Cope and Marsh bone-wars of the 1800s, researchers working in China usually found themselves picking sides as battle-lines were drawn. I was witnessing a landmark moment in diplomacy as well as science.

The fossils were like movie stars for us. There were 23 type specimens in the exhibition. A 'type', short for holotype, is the fossil that defines an entire species, the example against which all others are compared. These treasures of the Earth captured freeze-frames not only of a moment in time – the death of these precious animals – but of a moment in evolutionary history.

Once maligned by Owen as nothing but 'rat-like, shrew-like, forms', we now know Mesozoic mammals were the diverse pioneers of true mammaldom. With their unique synapsid inheritance of warmer blood, fur, milk and complex teeth, they blossomed in size, shape, ecology and numbers, playing a key role in the dinosaurian ecosystem. The story of their journey is told most vividly by unique fossils from China.

They are visually spectacular and scientifically invaluable. Laid out in life-like poses, these are the photo albums of our earliest truly mammalian relatives. Some have their fur intact, and the contents of their guts ready for inspection. Flipping through the pages of this deep time compilation, we search for ourselves, trying to understand the world we live in and who we are as mammals.

◐ ◇ ◑

The Jurassic was short, at a mere 56 million years but boy did it pack a punch. After their epic 250 million-year snuggle, the continents of Earth were parting ways. Laurasia was clambering off the top of Gondwana, and where they separated the sea rushed in to form the beginnings of the North Atlantic Ocean. The first cracks were showing in the south too, as India and Antarctica drifted away from Africa. By the end of the Jurassic, the whole world was unzipping itself, with new seas and oceans splashing into the gaps.

This increasing regionalisation of continents had consequences for the patterns of life on Earth. Alongside rocks, living groups were cleaved, allowing natural selection to work its independent magic on each one. Animals diversified in terms of number of species, and expanded the range of what they could do in their environments. We find the first members of lineages that are still around today: the ancestors of modern salamanders, birds and mammals. By the middle of the Jurassic animal diversity had exploded around the world, filling it with all the wonders that underpinned the next 100 million years, until the end-Cretaceous mass extinction.

The Jurassic also gave birth to China. For millions of years the pieces that would become this part of Asia were scattered like a paint splatter across the equator. They had fringed the eastern edge of the bountiful Palaeo-Tethys Ocean, sailing steadily north on their incremental journey through the millennia. By the Jurassic these pieces, in the form of the North China Plate, South China Plate and Tarim Plate, had come together to form what is now China. The orogeny of their meeting pushed sections of the plates underneath one another, and lifted others skyward.

Most of Europe, the seat of palaeontology at the dawn of Western science, was underwater in the Jurassic. China on the other hand stood proudly to the east, swathed in malachite forests and sparkling with freshwater lakes. Around two-thirds of China's coal reserves were formed at this time, the result of many millions of years of conifer, cycad and gingko woodland.

Conifers remain abundant globally to this day, but cycads are mostly confined to equatorial regions and botanical gardens. Rarer still, the gingko family has been reduced to just one living species, *Gingko biloba*, also called the maidenhair tree. It was once thought extinct in the wild, but a few trees still cling on in eastern China, and as ornamental additions to gardens and parks around the world. It is known for its distinct fan-shaped leaves, and it's an icon of China.

But in the Jurassic there were multiple species of gingko found across the northern hemisphere as a significant component of ecosystems. These gymnosperms were the descendants of the trees that replaced the lycopsid forests that had hosted our tetrapod origins in the Carboniferous. They included extinct species that evolved in unison with their insect pollinators, such as scorpion flies and lacewings. Proto-China was a rich landscape reverberating to the buzzing of insects.

China's forests were traversed by large reptiles, pushing through the foliage in search of fresh fodder: dinosaurs, large and small. Mammals scoured the branches and sniffed the undergrowth, while above them the skies were swept with pterosaurs. Fish and amphibians bred and flourished in the lakes.

We know all of this because an exceptional fossil record preserves it, plucked and pressed, like flowers in a book, by the Earth herself.

When tectonic shifts occur, the belly of our planet rumbles and bursts. The formation of China was no exception. Throughout the Middle to Late Jurassic periodic volcanic eruptions burst from the seams of the new country. Unlike the outpourings of the Siberian traps, these volcanoes were more explosive, and rained ash for hundreds of miles across the landscape.

Volcanic rock is not good for fossils, but volcanic ash is another story. In China, it acted like a preservative. The geological sequences are complicated, but in essence the fossil-bearing beds are a layered cake of alternating lake-sediments and ash. In times of peace, rocks grew slowly from the steady deposition of grains of mud and organic matter in rivers and lakes. Then a volcano or two would go off like a firework, sprinkling death over the landscape.

Ash is composed of very fine grains of minerals and glass that can be carried high into the atmosphere over long distances, coating everything they touch. It chokes waterways, clogs lungs, sticks to leaves and makes them inedible. Those organisms that perished in these mini-disasters were buried deep under a duvet of tuff. Beneath it, their bodies were safe from scavenging and so were left intact as a new cycle of sedimentation began. Repeat this. Repeat it again. Compress the lot under 160 million years or so of geological time, and then you have a snapshot of an entire Jurassic ecosystem.

A famous example of a similar occurrence in the modern world happened in Italy almost 2,000 years ago.[*] When Mount Vesuvius erupted in AD 79 the cities of Pompeii and Herculaneum were smothered in ash and pumice up to six metres (20ft) deep. Although many people escaped, on the second night a pyroclastic flow killed those who remained, cooking them instantly in a blast of gas and ash that reached 250° Celsius.

Archaeologists have dug up this world heritage site to reveal its secrets. The chilling moment in human history is captured by the ash layer, which hardened like a shell around the victims. The bodies of people, livestock and pets, as well as organic materials like wood, rotted away to leave voids – not unlike those of the Permian sand dunes of Clashach Quarry in Scotland. By filling these voids with plaster, the death poses of the city's inhabitants are revealed like gruesome musical statues.

Unlike Pompeii, the ecosystem of Jurassic China is preserved not in empty hollows, but complete with leaves, stems, wings, anthers,

---

[*] The 'ancient world' of classical Greece and Rome looks pretty modern to those who habitually work in millions of years.

feathers, bones and even fur. The fine-grained ash and sediments sealed their fallen bodies like sarcophagi, preventing decay and preserving them for eternity. They have *colour*. You can see what they ate for lunch. These aren't dusty relics, but fresh roadkill. As you lower your nose to peer at their skeletons, they appear so fresh, you can almost smell them.

◓ ◇ ◒

From their humble beginning in the Late Triassic, mammals soon made their home around the world. The last of the therapsids were long gone, and of the many cynodonts that proliferated after the end-Permian mass extinction, only mammals and their closest relatives, the tritylodontids, remained by the Middle Jurassic. They shared their world with a new fleet of reptiles on land, in the sea and in the air.

As those reptiles expanded and diversified, it appeared at first glance that mammals were overshadowed. In almost 200 years of fossil collecting, mammal remains from the Mesozoic were few and they told a somewhat dull story. It was wall-to-wall 'mice'[*] for 150 million years, so the story said. Dinosaurs oppressed and dominated them, preventing them from doing anything interesting. It seemed they were only good as lunch for the impressive beasts that surrounded them.

One of the first groups to emerge after the morganucodontans we met earlier, were the docodontans. The first fossil belonging to this group was found by Othniel Charles Marsh in 1880, and through the rest of that decade he discovered many more. He called it *Docodon*, meaning 'beam tooth'. From this animal the whole group was given its name.

Mammal fossils can be distinguished and named based on their teeth alone. Because of this, many of them are given names ending in 'don', from the Greek *dónti*, meaning tooth. This is possible because of the complexity of mammal dentition. The first mammals had stopped replacing their teeth multiple times, and instead grew one adult set of upper and lower molars that fit together well. After erupting the teeth could wear down slightly to occlude even more

---

[*] Of course they were not actually mice, because as we've already discussed, modern mammals like mice didn't exist at this point in time.

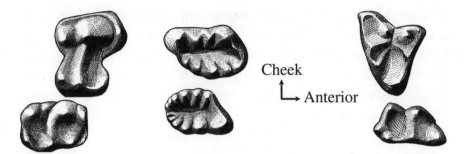

Figure 3 Occlusal views of molar teeth belonging to different mammal groups: docodontan (left), haramiyidan (centre), tribosphenic (right).

closely. These teeth were also developing into new shapes: raised ridges and cusps that meant they could chew insects more effectively.

In mammals today the most famous tooth shape is called tribosphenic. It's a complicated name for something quite simple. Think of teeth like tiny vistas: in the upper tribosphenic molar are three cusps, like mountains sitting in a triangle with steep gullies in between. The landscape of the lower molar is a bit more diverse, with three mountains on one side and a circular valley on the other, surrounded by foothills (usually three or more). There can be secondary peaks and troughs, or some peaks may be smaller or even disappear, but this is the basic shape of this iconic mammalian feature. By slotting the upper and lower teeth together during chewing, the tribosphenic molar allows mammals to shear and grind their food, like a mortar and pestle. The basic pattern has been adapted and modified in multiple groups of mammals, but they all come from this foundational blueprint for chomping.

The very first mammals in the Late Triassic and Early Jurassic, however, didn't have tribosphenic molars. Animals like *Morganucodon* had a little chain of mountains in a row – usually three or four – but not much else. It was a more complex set-up than their predecessors, but unfussy compared to their modern descendants. It also suggested that the earliest mammals ate a similar diet to one another, comprising mainly insects, and they certainly didn't have the grinding and shearing capabilities of later mammals. It was assumed that the simplicity of their teeth limited them ecologically. The first mammals appeared to have been simple shrew–like insectivores.

A study from 2014 helped change our thinking about the very earliest mammals' ecology. It was led by British palaeontologist Pamela Gill, and used fossils from *Morganucodon* and *Kuehneotherium*, both found in the Late Triassic deposits of Wales in the British Isles. These animals were extremely small – the epitome of mammal miniaturisation – and are known mainly from teeth and jaws. Gill and her co-authors were able to use these partial fossils to reconstruct a whole jaw for each species, in all its two-centimetre (less than one-inch) glory. By using calculations normally employed by engineers to test building materials, the team could assess how strong the jaws were along their length. The strength reflects how hard the animals could bite – and has implications for what they ate.

Although they appear superficially similar, it turned out that *Morganucodon* and *Kuehneotherium* had noticeably different biting patterns: *Kuehneotherium* had a less powerful bite, suggesting it ate softer insects than its compatriot. The team backed up this finding by looking at the microscopic scratch patterns on the tooth surfaces. They compared the scratches, known as microwear, to the patterns seen in the teeth of living bats that specialise in different foods. The microwear on *Kuehneotherium*'s teeth was more like that seen in the brown long-eared bat (*Plecotus auritus*), which eats softer insect prey. *Morganucodon*, on the other hand, had similar scratches to the teeth of bats that ate insects with harder outer coatings, like beetle carapaces. Even in these earliest of species with the simplest teeth, it seems that mammals were already differentiating ecologically.

The study of teeth is a huge part of mammal palaeontology. They are a unique identifier of different species, as well as telling us about diet. It is just as well that teeth are so distinctive, because most Mesozoic mammals are still known solely from those fossils and nothing else. There are more teeth in the fossil record than other body parts because they are composed of hard enamel and dentine, which resists the destructive processes of fossilisation. Their matchstick bones are rarely recovered, but hard teeth and dense jaw bone is more likely to survive.

As a mammal palaeontologist, I ought to be singing the praises of the humble gnasher. It's true that they still provide an incredible amount of information. They are also objects of beauty, gleaming up the microscope at you with their mountainous ridges and furrows. But I am willing to commit the ultimate sacrilege and say that studying teeth can be pretty dull. For much of the history of Mesozoic mammal palaeontology, it was *all* just teeth. They are tiny and far too complicated. Researchers have labelled them with a mind-bending array of occult terminology: trigonids and talonids, protocones and paracones, ectomesolophids and the enigmatic cristid obliqua. Nor are the same terms applicable to every group, or the same structures given the same names by every researcher. Learning the lingo is a painstaking process and often open to interpretation – indeed, passionate arguments continue over the identification of certain teeth and how they fit together during mastication. Lifelong enmities exist on the basis of a single bite.

Suffice to say I will not bore you with the mind-numbing minutiae of Mesozoic mammal molars. However, teeth still play a key role in our story. After Marsh discovered *Docodon* in the Late Jurassic sediments of the Morrison Formation in the United States, more docodontans were found at Jurassic sites in Europe and Russia. In the Jurassic they appear to have been widespread across what was once Laurasia, the northern half of the ancient world.

Docodontans are known by researchers as mammaliaforms. This distinguishes them from mammalians, the 'true' mammals. These terms are part of a revolution since the 1960s in the way scientists talk about animal taxonomy. It is called cladistics, and has replaced the traditional Linnaean hierarchical ranking system.

A clade is a group of organisms that all share a common ancestor. The clade Mammalia includes placental, marsupial and monotreme mammals and all of their relatives, back to their common ancestor around 160 million years ago.[*] These are the 'true' mammals. Anything outside of that grouping is not part of clade Mammalia, in the strict sense of the term.

The earliest mammals, including docodontans, are side branches that split off the tree before the common ancestor of mammalians. They therefore belong to the wider, slightly more inclusive category, Mammaliaformes. This group includes the clade Mammalia plus all the

---

[*] We don't know exactly when, and studies give different estimates, but it is *at least* this long.

other groups that share a common ancestor with the Late Triassic mammals, around 200 million years ago. Thinking of animal groups in this way, life becomes a series of Russian dolls, each clade nested inside the next. For synapsids, the biggest doll of all is Synapsida, holding all the animals discussed in this book within it, right back to the Carboniferous.

Researchers expected all mammaliaforms to be much simpler than their refined mammalian sisters. Despite attempting to shed the progressionist viewpoint inherited from the Victorians, people still often describe early mammaliaforms as 'primitive'. But as more docodontan teeth turned up, people were struck by the unexpected complexity of their cusps and ridges. Looking closely, it seemed that this very early branch of mammals had teeth that were almost a mirror image of a tribosphenic molar, quite literally. They had the same three peaks, the same pestle and mortar – the only difference was, they were in reverse positions compared to 'true' mammals.

Docodontans, despite not being mammalians, had evolved almost the same complexity in their teeth – and they did it first. These teeth in the mirror position to the modern tribosphenic molar were dubbed 'reverse' tribosphenic, or pseudo-tribosphenic molars.[*] This group was distinguished among the earliest mammal offshoots for possessing something so complex in structure. This undoubtedly made them more versatile, able to exploit a wider range of foods. Having said that, docodontans probably still ate mostly insects. Their complicated teeth were an interesting quirk of the group, but not exactly the kind of palaeontological news that sets the world on fire.

That is, until the first docodontan fossils were discovered from the Yanliao region in China. These docodontans were found not just with teeth and a few scattered bones, but with their entire skeletons intact, often in articulation. This modest group was catapulted from being a specialist's curiosity, to one of the most important clades in Mesozoic mammal palaeontology. What their little bodies revealed would transform everything we previously thought about Jurassic

---

[*] I find this weird, because looking at things chronologically, docodontans came up with the tribosphenic cusp pattern first. So shouldn't modern mammal teeth be the 'pseudo' ones? Of course it doesn't work out that way, because we look backwards from the present.

mammals. It is in docodontans that we first see the true power of teeth to unlock mammal potential.

Outside the Beijing Museum of Natural History, the security guard squinted at me as I waited for clearance. My telegraph-pole stature was a source of endless curiosity in China. I stood on the dusty pavement at the security gate with a small mix of scientists from the United States, Russia and Germany. Three members of the group were late-middle-aged men, the established experts in Mesozoic mammal palaeontology, and two of us were young women early in our careers (me and Dr Simone Hoffman). We made a visually interesting motley assortment.

The director of the Museum, Qing-Jin Meng, met us in the wide forecourt. He was a key organiser of the Symposium of Mesozoic Mammals, along with my PhD supervisor and collaborator, Zhe-Xi Luo. A well-built man with a healthy shock of black hair, Meng was a generous and welcoming host. He extended a strong handshake, then led us to a door at the back of the building. Inside there were refreshments in the anteroom to help us recover from the intense heat on our journey through the streets. But we shuffled our feet with impatience: this morning we were due to meet our Chinese Mesozoic mammal superheroes.

In 1999, some 400 kilometres north-east of Beijing, farmers digging their land for planting found the most spectacular Jurassic mammal fossils in the world. The landscape around their village, Daohugou, is hilly and green. It lies in the southern edge of the great grasslands of the Inner Mongolian plateau, interspersed with forests of sweet-smelling pine. This part of the People's Republic of China, the Inner Mongolia Autonomous Region, is home to five million Mongols and about 20 million Han Chinese. The region cups the Republic of Mongolia to the north, and covers an area of over a million square kilometres stretching east to west along the top of China. With the long hard winter over, the farmers of Daohugou were preparing their land before the hot, humid summer took hold.

Daohugou village itself isn't on Google maps. It lies in a loop of Inner Mongolia hanging down between Hebei and Liaoning Provinces, north of the Yan Mountains (Yanshan). This whole region is called Yanliao. To find Daohugou you must turn to the scientific literature. Thanks to the fossils found there it has become famous among researchers, and is very much on the palaeontological map, if not the sat nav.

The farmers had found the first fossils near Daohugou village in 1998. In the summer of the following year two more fossil skeletons were recovered, belonging to salamanders. They heralded the first drips of a flood of discoveries from this region and other outcrops in neighbouring Hebei and Liaoning Provinces. At the last count, these fossils included over 350 species of insect, 87 kinds of plants, 15 pterosaurs, at least 12 kinds of mammals, 8 different dinosaurs, 6 species of amphibians, and an assortment of fish, lizards and aquatic invertebrates. This treasure trove has been dubbed the Yanliao Biota, and it is one of the most exceptionally preserved insights into the deep past recorded anywhere in the world.

We were about to see these fossils in the Beijing Museum of Natural History, where they had been brought for the Mesozoic mammal conference. In the laboratory space beyond, a series of microscopes had been set up for us. It was a windowless room, but bright. The air was cool, making my bare arms prickle, but the warm beam of lamps lit the cloud-grey work surfaces, and made the wooden table in the centre of the room glow like amber.

I approached, holding my breath. In front of me were two palm-sized, flat rocks. They held two halves of one animal, pressed between their mustardy pages. It was small enough to put in your pocket, with one back leg sticking straight out from the bottom. The skeleton was surrounded by a dark smudge that appeared to be the remnants of a round fuzzy body. The head and shoulders were in profile, and it looked for all the world like a tiny Nosferatu tip-toeing across the stone surface with claws outstretched. I squealed a little as I realised what I was looking at: this was the Jurassic mole, *Docofossor*.

The tiny creature had turned to stone, but it felt light in my hand as I gingerly lifted it. Seeing what I was doing, Zhe-Xi Luo joined me at the table. He pushed his glasses closer to his round face and beamed at me. 'Isn't it spectacular?' We placed the tiny body under the microscope. The consummate teacher, Luo talked me through what I was seeing.

The skull was snubbier than other docodontans, and the teeth reduced in complexity and similar in shape to modern mammals that forage for food from below ground, like worms. The limb bones were proportionally short, but it had long sharp elbows (called olecranon).

These provided lots of attachment area for powerful upper arm muscles to dig through soil. All of these features hinted at the kind of creature we were looking at, but there was one further attribute that set it apart. Luo and I dwelled upon its most impressive front feet.

Luo was one of the first people to examine *Docofossor*. Although based at the University of Chicago, he collaborates with researchers at the BMNH. This specimen lay in their collections, one of many mammal fossils from the Yanliao Biota. Peering down a microscope not long after its discovery, he could hardly believe what he was seeing. This creature had adaptations never found in another Mesozoic mammal. 'It was missing one digit segment on each finger,' he told me. 'I saw that and I thought, wow, we really have something special here!'

What Luo had found was nature's first mole paws.

◦ ◊ ◖

Moles in the modern world include animals from completely separate families. They aren't closely related, but they have one thing in common: they have shovel-shaped paws.

When most people talk about moles, they mean members of Talpidae. This family is most closely related to shrews and hedgehogs. These scourges of the perfect lawn include the European mole (*Talpa europaea*), the short-faced mole in Asia (*Scaptochirus moschatus*), and the eastern mole (*Scalopus aquaticus*) in North America. They have liquorice-black to chocolate-brown velvety fur, chubby round bodies with short limbs, and pointed noses. This family is made up of specialist diggers and their body plan is modified to the extreme for a subterranean lifestyle. Their front feet are wide with long, thick claws, ideal for shifting dirt to make tunnels and find food. They have no external ears, and have tiny eyes – sight, after all, is not very important to them. Touch, on the other hand, is vital, and in none more so than the star-nosed mole (*Condylura cristata*). This little digger has a frilly rim of 22 finger-like projections around the nose, containing over 25,000 touch-sensitive Eimer's organs. Many moles have these organs, which are modified skin cells packed with nerves, but *Condylura* has gone all-out, investing in developing them to the extreme. The result looks somewhat freakish to our sensibilities, but they work with such delicacy that they can find food and thrive by feeling their way in complete darkness underground.

Being mole-like is not confined to the talpids. Two other groups of modern mammal include members that have convergently developed the same adaptations. The golden moles, Chrysochloridae, are common in southern Africa and are related to other animals of that continent, such as elephants and aardvarks. Meanwhile in Australia an even more distant cousin has also turned to moledom: the marsupial mole *Notoryctes*. Like its above-ground kin, *Notoryctes* has a pouch in which it raises its young, which opens to the back to prevent it from filling with sand. But there the resemblance ends, because everything else about this animal screams of the underground.

These three groups have all become mole-like in similar distinctive ways, despite sitting at entirely different tips of the mammal tree. Natural selection has acted on them to solve the same ecological problem, and in each case has come up with the same results.

Hold up your hand, palm facing away from you. Anatomists have allocated your fingers ascending numbers: your thumb is number one, your pinkie is number five. This numbering system is universal in all tetrapods because we share a five-fingered body plan dating back to our common ancestor. As animals have diverged and specialised for different environments and lifestyles, this basic plan has been altered and modified – a kind of evolutionary recycling. But limbs and digits are homologous, a fancy way of saying that they share the same evolutionary origin, even if their function changes. So whatever happens, your middle finger will always be number three, even if you lose some fingers through natural selection, mutation or accident.

Like most other mammals, you have long metacarpals (the bones inside your palm), then three phalanges on each digit (except your thumb, which only has two). Again, the basic pattern of your finger bones is homologous, but can be modified through natural selection.

One of the groups of moles, the African chrysochlorids, have done just that. Taking their adaptations to life underground a step further, golden moles have not only widened their front paws and strengthened their claws, they have also reduced the number of fingers and the number of bones in each finger.

The golden mole loses its phalanges in the womb. In the earliest days of embryonic development it has five fingers and three phalanges just like any other mammal. But as the embryo continues to grow, the two middle phalanges stop growing and begin to merge. By the time the animal is born it has just two phalanges in most of the fingers, and the pinkie is gone altogether. The fingers are reduced in size, and in some species the enormous clawed middle finger dominates completely, effectively giving them pick-axes for hands.

It was thought that being a mole was a relatively recent specialism. We had counted moles alongside all of the other highly specialised mammals – tree-dwellers, swimmers, gliders and flyers – as animals that could only exist in the modern 'age of mammals'. Scientists came to this conclusion because they assumed mammals only radiated and explored new niches after the disappearance of the dinosaurs. Until the turn of the century, no fossil evidence existed to the contrary.

Then *Docofossor* came along. Examining those tiny shovel-paws down the microscope, Luo saw they had just two phalanges in each finger, whereas other docodontans have three. This was significant not only because it proved this animal possessed the first suite of highly specialised mole-like adaptations in the fossil record, but also because of what it tells us about embryonic development and gene expression in our earliest relatives.

The medical term for shortening of the finger bones is brachydactyly, literally meaning 'short finger'. It can happen to animals (including humans) as a random mutation or part of particular medical conditions. Through genetic studies on mice, researchers have identified the gene-signalling pathways that control the formation of the limb during the development of an embryo. A failure in one of the development pathways leads to brachydactyly in mice and other mammals. This means *Docofossor* provides one of the rarest pieces of information: an insight into gene signalling and development across geological time.[1] By implication, a host of changes in fossil mammal physiology and anatomy could be traced to genes identified in modern developmental biology.

Using fossils like this, we are pulling back the curtain to see more of the mechanisms of evolution in deep time.

This short-fingered Jurassic mole would be extraordinary enough, but just wait until you meet her sisters. Appearing in the same issue of the scientific journal *Nature* as *Docofossor* was yet another specialist. In the treetops of Jurassic China, *Agilodocodon* was gracefully exploring life in the canopy.

This docodontan took an opposite route to her sister, not only in terms of chosen hangout, but also physically. It was slender-limbed and long-fingered, equipped to grip tree-bark and branches. *Agilodocodon* had the narrow waist of a gymnast. Mammaliaforms had already reduced and lost the ribs below the chest, which we saw in previous chapters provided flexibility and tied in with the development of the diaphragm. In *Agilodocodon* this reduction was even more distinct, freeing it for the kind of high-wire contortions necessary for life in the trees.

And that's not all. The first mammal fossil described from the Yanliao Biota back in 2006 showed even more unexpected specialisms – for living not underground, but underwater. The name of this animal, *Castorocauda lutrasimilis*, seriously gives the game away:[*] it means 'beaver-tailed and otter-like'. It was another docodontan, and another example of this group's surprising diversity.

Castorocauda is known from an almost complete skeleton that lies with legs akimbo on its stone-slab grave.[2] The true glory of this specimen is that it is preserved with soft-tissue impressions surrounding it, like a fluffy aura. It has an especially fuzzy bum, and hairs can be traced clearly fringing its whole pelt except for the hands, feet and tail. It is the back end that is most revealing: the outline of the hairless tail is wide compared to the body. This told researchers that it was broad and paddle-like, more like a beaver's tail than that of any other docodontan – or any other Mesozoic mammal known. Inside it, the tail bones themselves had expanded sideways, just like those in semi-aquatic mammals known today that use their tails to help propel them through water.

Taking a second look at the paws, *Castorocauda* appears to have had webbed feet. Those complex docodontan teeth had been simplified slightly and curved backwards like fish hooks. The ribs were expanded

---

[*] Which is obviously the whole point. A well-chosen scientific name should traditionally let a person know immediately what the defining features of an animal are, but you had to have a classical Western education to understand it. Although done for a logical reason, this is one of the many biases of traditional scientific practice. These days, however, a scientific name is just as likely to tell you which *Harry Potter* character the researchers prefer.

into strips rather than rods, stiffening the body. What's more it was big. Whereas other docodontans were no bigger than a well-fed hamster (around 100g at best, or 3.5oz), *Castorocauda* would have weighed in at around 800g (almost 2lb). That's over 30 times more than the stereotypical 'shrew-like' early mammal everyone thought they knew.

Castorocauda was more like a platypus than a mouse. To their amazement, scientists had found in the Jurassic a substantial-sized animal, well established in a semi-aquatic niche. It was able to move with a swift tail pump and manoeuvre its robust body using paddle-feet, and to catch slippery underwater insects – perhaps even fish – with sharp teeth.

The docodontans were busting all the Mesozoic mammal myths. Researchers soon realised they were not the only ones. A second group of creatures was also breaking new ground.

In the same year that *Castorocauda* was announced, the fossil of *Volaticotherium* caused a stir. This Chinese specimen is less attractive than the docodontans, with bones in disarray, but its promise lay in the preservation of soft tissue. A dark oval of seemingly superfluous skin darkened the rock next to the skeleton. Looking closely, the team who described it, led by Jin Meng from IVPP and the American Museum of Natural History, realised this splodge was the remnant of a patagium, a flap of skin stretching from fore to hind feet.[3] This was the first ever *gliding* mammal.

Gliding is not the same as flying. Bats are the only mammals to have evolved true, *powered* flight. Despite their name, what today's flying squirrels do is actually gliding.* There are a number of mammals that employ skin flaps to travel through the air: two groups of squirrel, the colugo and some lemurs, marsupials like the phalangers, feather-tailed possums and appropriately named gliders (like the sugar glider, *Petaurus breviceps*). Reptiles have also got in on the action, as have amphibians

---

* Insects were the evolutionary Wright brothers, and among tetrapods it was the pterosaurs who figured flying out a few hundred million years later. Birds and bats are the only powered flyers among modern tetrapods. We now know how birds emerged from dinosaurs, but the origin of bats remains a mystery. The first bat fossils are about 52 million years old; they were already able to echolocate and were 100 per cent batty in form. Hopefully one day someone will find a proto-bat, and the mystery of their evolution will become clear.

and even fish, each finding ways to modify their body plan to make an aerofoil and exploit the properties of lift. It can be extremely effective, taking animals hundreds of metres between tree trunks to escape predation or reach new territory in search of food and shelter.

Since the announcement of *Volaticotherium*, many more gliding Jurassic mammals have been discovered. Some, like *Maiopatagium*, look as if they are frozen in the act of falling from the sky, with their skin membranes billowing around them like a parachute. Although none of these skin and fur impressions suggest Mesozoic mammals were brightly patterned – most modern mammals aren't either – they provide a startling insight into the lives of our ancient kith* and kin.

Many of these gliders belong to a prolific yet enigmatic group named the haramiyidans. These animals are known for their strange molar teeth, comprising a long V-shaped valley between two rows of rounded mountains. Meanwhile, their incisors were large and jutted out at the front, like those of the modern rodent. These teeth are probably adaptations for a plant-based diet, or perhaps omnivory. It is unclear exactly where they fit into the mammal family album, but their Chinese fossils push back the earliest record of gliding mammals by over 120 million years, adding yet another string to the ecological bow.

Finding these specialised niches in mammals of the Late Jurassic has rewritten the textbooks. Around 160 million years ago in a world replete with giant reptiles, docodontans tried their hand at almost everything. Not only did these Chinese fossils overturn our understanding of the emergence of ecological diversity in mammals, but they did so in a group that wasn't even technically part of Mammalia. These ground-breaking mammaliaforms were not cowering in anyone's shadow; they were feasting on fish, exploring new heights and tunnelling to victory.

It is thought that the key to unlocking diversity in this humble mammal branch lay in their complex teeth. Being able to chow down on a wide range of foods may have meant that the first docodontans

---

* 'Kith' meaning friends and neighbours rather than family – as I've outlined previously, we are probably not directly descended from most of the animals in this book.

could seek out new life and new niche-space* that remained inaccessible to their compatriots. By hitting upon the prototype tribosphenic molar, docodontans underwent a trial run of future mammal potential.

Haramiyidans probably found their unique place by exploiting more plant foods than other groups. The forces of natural selection would have acted on them to enhance this advantage, resulting in changes across the skeleton. Although we can't say for certain that they all lived at exactly the same time (the rock record has a resolution counted in the hundreds of thousands to millions of years), it is nevertheless shocking to find them all in the Yanliao Biota together. The reason for this diversity is not that China held some special mammalian elixir – there is no evidence that the ecosystem was unusual for the time – it is just that the fossil record there is so exceptional. The inhabitants of ancient China belonged to widespread groups, most of them represented at other sites in the northern hemisphere at least, if not also in the southern. In most other places, however, they are known from much less complete skeletons.

To find out if mammals in the rest of the world were similarly adventurous we need to find their tell-tale bones. Hands and feet, arms, tails, backbones – all of them alter as organisms adapt to the demands of an artisanal niche. What China's Jurassic treasure-chest contains is a promise: that the time of dinosaurs was no one-person show. The incredible showcase of mammal diversity we see around us today has deeper roots than we had ever imagined.

To understand the beginning of docodontan diversity, we need to see what the earliest members of the group were getting up to. Some of the answers may lie entombed in the sea-swept coastline of Scotland.

The sun was high, the tide was low. I found myself at the edge of the shore. The marmite kelp was exposed like toes as the blanket of the sea pulled back. I was a limpet, sitting on rocks that didn't feel the kiss of air for most of the tide cycle. I could smell the steamy pickle of the deep rising around me. I crawled like a crab with my face against the stone, scuttling in pursuit of bones.

---

* Yes, it's a *Star Trek* reference. Don't judge me.

Loch Scavaig was holding its breath. It was a twin of the cloudless sky, both turned blue. Across the water were *An Cuiltheann*, the Black Cuillin – Skye's iconic ridge of ragged mountain peaks. Sgùrr Alasdair, Sgùrr Dearg, Sgùrr na Banachdaich. They were still wearing woolly jumpers of snow. Like any Islander they knew the weather was fickle – for all the blaze of our nearest star just now, if the wind had a change of heart it could still blow in fearsome March storms. Best to stay sensibly clad.

Despite this, my own jumper was discarded up-shore with the rest of our equipment. The afternoon was hot enough to pop seaweed. I lifted my head and squinted along the seafront where my team had spread out in search of fossils. Occasionally one would meerkat up from behind a boulder, take a few steps, then drop out of sight again.

The rocks tell stories of a very different Skye. They belong to a series of strata that formed in the Hebrides basin in the Middle Jurassic, called the Great Estuarine Group. These rocks raise their heads on the Islands of Skye, Raasay, Muck, Eigg and Rùm. They include sandstones and limestones, mostly from the bottom of fresh lagoons and salty shorelines, sometimes from the bed of shallow seas and tidal pools. They are the most fossiliferous Mesozoic rocks in Scotland.

The section my team works on is different from the rest. These limestones are from a warm Jurassic lagoon. It was nothing like the cold brine of Loch Scavaig. These still pools were fresh and sweet, brimming with crocodiles, fish and aquatic reptiles. The layers in the cliffs charted cycles where drier conditions evaporated the water to dust and the surface was parched and cracked for a time, before refilling. Had I been dipping my toes in that fresh water, there would be no Black Cuillin to gaze at in repose: the fabric of those iconic peaks wouldn't erupt for another 100 million years. Instead, behind me the Scottish mainland – a chunk of stone formed when the Earth was young – would stand proud and high on the horizon, coated in conifer forests.

Although the fossils from Skye are not as exceptional as the Yanliao Biota, they have many things in common with their Chinese counterparts. Both are Jurassic in age, although the Scottish fossils are slightly older. Both sets of fossils are from freshwater environments. Both have a similar cast of characters living in and around their shores. And both sets of fossils comprise more than just a few teeth. There is enough at each to reconstruct a portion of their complex ecosystem, as well as of the animals themselves.

The jaw that Michael Waldman found in Scotland in the 1970s belonged to a docodontan. He and Savage named it *Borealestes serendipitus*, 'the northern rogue found by luck' (although it truth, it was more like advance planning by Waldman). It was the first Mesozoic mammal found in Scotland, but far from the last.

Multiple species were already known from the Stonesfield Slate in England, which had produced the very first Mesozoic mammals to be described in the early 1800s. A handful of other localities had cropped up in the subsequent 150 years. In the 1970s one locality became key for our knowledge of Jurassic mammals in the British Isles: Kirtlington Cement Quarry in Oxfordshire.

Kirtlington had produced a few bones in the 1920s when it was worked commercially. But it was not until amateur geologist Eric Freeman prospected it 50 years later that anyone realised the potential for finding small vertebrate remains. First Freeman, then a team from University College London, worked what became known as 'the Mammal Beds'. They dissolved the rocks and sieved the dregs, then picked through the remaining sediment to remove bones and teeth. A tedious method, but one that proved wildly successful.

There are hundreds of fossils from Kirtlington. They form the core of the English Mesozoic mammal collections at the Oxford University Museum of Natural History, and the Natural History Museum in London. Among them are multiple species of docodontan, as well as other mammal groups – including some of the earliest members of Mammalia. They are found alongside lizard, amphibian, fish and turtle fragments, pterosaur teeth and pieces of dinosaur. Since the 1970s this rich Middle Jurassic site has been regarded as the best in the British Isles for such fossils, and among the most important in the world for the range of animals it captured.

While sieving is an excellent approach for recovering many different species, it does so in fragments. Few of the animals identified from the Oxfordshire beds comprise more than a tooth or two. Random bones emerge jumbled like woodchips. With no corresponding bodies or teeth to match them with, they can't be identified as anything other than indeterminate parts.

Scotland is different. The mammals and other small vertebrates from Skye are mostly found in bundles. They are tiny and unpromising to behold – many look more like seagull droppings than fossils. But with the revolution in the accessibility of CT scanning our team is able to study these ugly blobs. To our astonishment it turns out that they comprise multiple bones, sometimes near-complete skeletons. Often, the uglier the specimen, the more impressive it turns out to be when scanned.

They are a far cry from the hieroglyphic Chinese specimens, pressed with limbs outstretched between the sheets of time. But the advantage is that these Scottish bones aren't squashed flat; they retain their three dimensions, capturing their true shape. Getting them out of the rock is a challenge, as Waldman and Savage had discovered. The limestone is baked hard by the volcanic eruptions that later formed the Cuillin, and it doesn't respond well to acid. There can be no bulk sampling on Skye as there has been in Oxfordshire. But the trade-off is fossil completeness to rival the Yanliao.

Back in the office at National Museums Scotland, I worked through my synchrotron scan data from France. It revealed that the skeletons from Skye were docodontans, those trail-blazers of the Jurassic. By comparing them to the stunning specimens from China, it seemed our British animals were closely related, probably from earlier in the family tree. This meant they might represent the ancestral body plan for the group.

My next task was to write a painstaking description of their skeletons, and to search for any signs that these Jurassic islanders were running their own version of the ecological *Wacky Races*.

◖ ◆ ◗

Understanding the ecology of a species – the way in which they interact with their environment – is of great interest to palaeontologists. Since the discovery of unexpected docodontan diversity, it has become a particular preoccupation of Mesozoic mammal researchers. One of the main ways to explore the subject in fossils is to examine their ecomorphology. This is the way in which their bodies are shaped by their habits; in other words, how their form relates to their function.

Whereas ecomorphology is an old concept, the way we explore it has changed a great deal. From the original painstaking comparisons of Cuvier and Owen, we now employ coding and computerisation to add statistical rigour to the process. Sometimes observation can be deceptive. The goal is to test observations to make sure they are mathematically significant.

For example, it may be true to say that bears have a particular ankle shape because they walk with a distinctive ambling gait. You might conclude that the ankle shape was driven by this movement, that the form of the bear ankle is ecomorphologically connected to its function. But how can you be sure that the ankle shape of bears isn't really the result of the fact that all bears are closely related to one another? They may inherit the shape simply because they are in the same family. To try and sort out this problem, researchers have to use programs to statistically test their data. They test the relationship between form, function and phylogeny (animal relationships). If the relationship between form and function is stronger than between form and phylogeny, it's safe to assume that they are looking at ecomorphological specialisations rather than a quirk of that animal group. In the case of bears, although they certainly walk with their own unique amble, their inherited ankle shape is so odd that it obscures the ecomorphological signal. For bears, it seems their phylogeny plays a greater role in ankle shape than their ecology.

The length of limb bones is especially revealing when it comes to animal movement. We saw earlier how the long fingers of *Agilidocodon* told us it was a capable tree-climber, whereas the brachydactyly in *Docofossor*'s shovel-paws betrayed it as a mole-like digger. A foundational example of the role of limb proportions in animal movement is the length of legs compared to the body.

Running species have longer, more slender limbs, especially the extremities. Horses, cheetahs and sengis* are good examples of this. Animals that don't walk as much or don't habitually run to hunt prey or evade predators tend to have shorter limbs by comparison – think of rhinos, wombats, or at the extreme, sea lions and other animals that can

---

* Also known as elephant shrews (Macroscelididae), these little critters are some of nature's triumphs. They are no bigger than a small hedgehog, but can run at almost 20mph (30kph). That would be like a human running at 100mph, four times the speed of the Olympic athlete Usain Bolt.

barely walk at all. These differences feel intuitive, but in reality the answers can be complicated by the multiple functions expected from limbs and other bones in the body. Among some animal groups the differences can be so small that they are hard to detect. Length and width measurements are useful to a point, but they don't capture complicated shapes like joints, nor do they work well in three dimensions.

One of the most common ways modern researchers test form versus function is through something called geometric morphometrics. This method can be done in two dimensions (using photographs of bones) or three (using CT-scan data to make models of bones). In either method, points are identified on the bone that together capture its overall shape. These are called landmarks, and they are a bit like motion-capture film, as used in special effects or animation. By placing landmarks in the appropriate places you can mark the parts of the bone that are important. To analyse shape using geometric morphometrics, you identify the same landmarks on lots of different bones, then compare the distribution of these landmarks throughout your whole dataset. How do the same landmarks in the ankle of a bear and a cheetah compare?

The results of geometric morphometrics can be placed on a kind of chart called a morphospace. The position an individual animal's datapoint occupies in the morphospace should provide information about the shape of their bones compared to other animals. In this case it will hopefully reflect their ecology, so that all of the specialist digging animals cluster together on the chart, whereas those that habitually climb and live in trees will cluster together elsewhere. A bear and a cheetah should be placed far apart no matter what bone you are testing.

As well as assessing the results of geometric morphometrics visually, you can test them statistically. If you do this analysis using a dataset of modern animals, you can then add a fossil to see where it plots on the morphospace. Hopefully this will tell you what living animals the fossil most closely resembles, whether their similarity is statistically relevant and actually reveals something about their ecology.

Geometric morphometrics are now being deployed on Mesozoic mammals and kin to test their true ecological diversity. But there are some snags when analysing animals that lie this far back in time. The first is the lack of complete fossils. For all the hundreds of species named, most are only known from teeth and jaws. This can tell us a bit about diet, but not about locomotion. For the fossils that include limb bones, those bones are often damaged, and in the case of the beautiful roadkill from China they are also squashed. This makes them difficult to CT-scan, and compresses the features researchers want to use for their analyses.

But perhaps the most challenging issue is that the animals themselves are separated by hundreds of millions of years of evolutionary change. Superficially, we might think Mesozoic mammals resemble modern mammals. They look like rodents to the untrained eye – their bones, however, tell a more complicated story. Many of them are a mixture of shapes inherited from their predecessors, alongside new features.

Going back to the main ankle bone for example, called the calcaneus – a bone that tells us a lot about foot posture and locomotion in modern groups – it barely changes between the Triassic cynodonts and the inhabitants of the Late Jurassic and Early Cretaceous. The pectoral girdle that gave rigid upper-body support to our synapsid ancestors with their sprawled limbs, shrinks, but only properly disappears in the Cretaceous. The monotremes, just to be awkward, kept theirs, making them unique among living mammal groups.

Meanwhile, completely new bone shapes evolved as muscles shifted to perform new tasks and body mass changed. For small-bodied animals like the Triassic mammals, there are additional problems for researchers. Small animals don't need to adapt their skeletons very much to move in new ways because they are light enough to climb or scoot using existing musculature alone. This means the ecomorphological signals in their bones are often harder to detect.

For all of these reasons direct comparisons between a living animal and its long-lost cousins may be meaningless or hard to interpret, because their bodies just don't work the same way. These are the caveats palae-ontologists spend their careers accommodating. They present challenges, but not insurmountable ones. With each new piece of evidence, however small, our picture of ancient life is clarified and refined. Sometimes, as in the case of Mesozoic mammals, it is transformed.

The Jurassic is significant for us because it is also when the earliest of our modern mammal lineages appear. We commonly call the two main mammal houses alive today marsupials and placentals. However there are more technical names for them that encompass extinct side-shoots from their Mesozoic days. Metatherians are the marsupials and all of the extinct mammal groups most closely related to them. Their sister group are the eutherians, which comprise placental mammals and all of the mammals more closely related to them than to metatherians. Together, these comprise the therian mammals, the stock to which we and all mammals on Earth today, except monotremes, belong.

There is some debate about when eutherians and metatherians went their separate ways, but the ancestors of therian mammals certainly lived in the Jurassic. The first mammal fossil ever described, that modest 'opossum-like' jaw bone from England, was undoubtedly from near the base of the therian tree.

One controversial contender for the earliest eutherian comes from the Yanliao deposits. It is a mammal named *Juramaia*, a truly tiny little creature living up to the old cliché of the 'shrew-like' ancestor. Only the front half of the animal remains flattened on the slab with arms outstretched in a cartoonish splat. However not everyone agrees that it is eutherian. It is not until the Cretaceous that we can be sure our two modern mammal lineages were forging their independent evolutionary paths.

Around 160 million years ago however, therians were nothing special. Their teeth were complex, it's true, but in terms of ecological innovation their bodies don't suggest they were joining in the ecomorphological party taking place around them.

But the innovations of Mesozoic mammals go beyond teeth and limbs. One biological rearrangement in particular would hone mammal ears into something unheard-of in any other tetrapod group. Although the Jurassic was undoubtedly a time of thriving mammaliaforms, the true mammals were also evolving novel new anatomy.

◦ ◇ ◌

Early mammal fossils are especially crucial for what they can tell us about the changes taking place in one key area of the mammal skull: the ear. The ear is the one part of the mammal body that has been scrutinised as closely as teeth. There is a good reason for this, which we first touched upon in Chapter 7. But there is much more to the story.

Modern mammals can hear sound at ranges far beyond other tetrapods. Sound frequencies are measured in hertz, and humans can hear between 20 and 20,000 hertz. We lose the ability to hear higher frequencies as we age because we lose the fine 'hairs' in our cochleas. Even at our youthful best, human hearing is rather mediocre among mammals, neither especially high nor low by comparison. But among our brethren, there are some who have taken hearing to piercing heights.

At the top end of the spectrum, the highest frequencies are detected by bats and dolphins. Some bats can hear as high as 200 kilohertz (that's 200,000 hertz), and dolphins reach as much 275 kilohertz. For marine mammals the hearing range can also stretch to the opposite extreme: blue whales communicate across entire oceans with low rumbling sounds they pick up below 10 hertz. On land, elephants feel sound over vast distances through their feet, and can detect vibrations as low as just 14 hertz.

These incredible audio feats are only possible thanks to natural selection pimping mammal ears from their tetrapod body plan. Because of the impact this super-sense has had on mammal behaviour and ecology – changing how they hunt and communicate – researchers have sought to unpick how ears evolved. As they traced this transformation through the mammal lineage, they discovered a textbook example of waste-not-want-not in our evolutionary history, but the pattern it took was unexpected.

Sound is basically vibrations in the air. It reaches our outer ear first, the external pinnae or earlobe that funnels sound and helps us pinpoint the direction it came from. Unlike most mammals with pinnae, ours are no longer mobile – even the most impressive human ear-wigglers aren't a patch on the swivel your cat can achieve in response to the faintest rustle of a kibble packet.

The sound travels down the ear canal and bangs the eardrum. This is the beginning of the middle ear. In mammals the middle ear has three tiny bones in it which transfer sound to the inner ear. These bones are the malleus, incus and stapes, and they are the smallest bones in the body. They don't just relay messages from the outside, they magnify them. Whereas other tetrapods have a stapes, only mammals have added an extra two bones, massively improving the magnification of sounds. The effect is similar to the difference between hitting a ball with the palm of your hand or with a golf club. The ball is hit harder and travels further with a nice long wallop with a 9 iron.[*]

Once inside the inner ear the sound travels through the fluid of the cochlea. This liquid, called endolymph, carried the vibrations to tickle tiny 'hairs' on the inner surface of the cochlea. The tickles are turned into electrical signals and sent to the brain for interpretation, like Morse code to a telegraphist. The shape of the inner ear, particularly the length and curvature of the cochlea, can tell us a great deal about how well an animal hears. In the fossil record it can also give clues to family relationships, as the nerves and blood vessels that pass through the bones housing the ear change not only through time, but also between lineages.

The first synapsids had a jaw joint where the quadrate in the skull met the articular in the jaw. This arrangement is ancestral to all jawed vertebrates, including fish. They also had multiple bones comprising the jaw, including the dentary, angular, surangular, prearticular and coronoid. By the time we get to cynodonts however, the jaw was primarily made up of the tooth-bearing dentary bone, and the others were lost or reduced and sent to the back.

The reduced jaw bones were not useless however, playing a role in transmitting sound to the existing simple ear through vibrations in the jaw itself. One bone, called the angular, became plate-like to reverberate more effectively and send signals into the cochlea via the stapes. The quadrate and quadratojugal in the skull were also involved in sound transmission, but their tight coupling – which reinforced the jaw joint – limited how much they could move. This arrangement, known as the mandibular middle ear of cynodonts, or MMEC, is our starting point.

Getting from the ancestral MMEC to the definitive modern mammal ear, or DMME, is one of the miracles of evolutionary transformation. The bones of the jaw reduced further – partly because modifications in chewing drove changes in the muscle

---

[*] I'm not a golfer, but I'm told a driver or 3 wood achieves the greatest distance. It doesn't sound as good though.

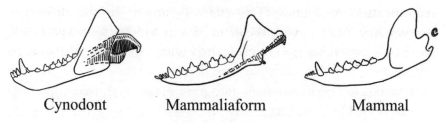

Cynodont   Mammaliaform   Mammal

Figure 4 The shrinking jaw bones and their incorporation into the middle ear through mammal evolution.

arrangements, freeing up the bones from the responsibility of carrying muscular loads. The quadratojugal is among the first to go, allowing the quadrate more movement. Everything at the back of the jaw started to shrink, but because the bones still played a role in sound transmission they were not lost. In mammaliaforms such as the docodontans, there is still a wide trough along the back of the dentary where these bones nestled, called a postdentary trough. The prearticular became cartilaginous, and leaves a distinct trace on the jaw called a Meckel's groove where it snakes along the inside of the jaw bone.

So far so good, but how and when did we stuff all these bits into our ears? As we saw in the Late Triassic, the very first mammals were all small and nocturnal. Their reduction in size was a crucial precursor to the development of mammal hearing, freeing their postdentary bones from their jaw joint completely. It appears mammals may have developed modern ears and enhanced hearing to help navigate in their nocturnal world. Sounds straightforward, but it turns out the picture is more complicated than a simple shrinking of the bones through time.

Monotremes, as always, totally mess things up. Their ancestral group split from the rest of the mammals at least 160 million years ago. We know this because the very earliest fossils from their lineage date from that time: animals like *Shuotherium*, found in China and England, or *Henosferus* from Argentina. Later in the Cretaceous, the earliest monotreme *Steropodon* appears in the rocks of Lightning Ridge in Australia. But all of these fossils still have a Meckel's groove. They didn't have a DMME, even though their modern descendants do.

Things get even more complicated in the rest of the mammals. The precursors to therian mammals don't have a modern mammal ear, but the multituberculates – a closely related group of rodent-like animals that branch off the mammal tree before them – do. Then the metatherian and eutherian mammals have them, seemingly appearing from non-DMME ancestry.

What on Earth was going on? Palaeontologists had expected to trace a nice neat evolutionary line from cynodont to modern mammal ears, but it seemed from the fossil record that mammals were flipping backwards and forwards. This was incredibly unlikely – you can't 'evolve backwards' – such reversals are exceptionally rare, if they even occur at all. There had to be another explanation.

An answer to the riddle of the ears was outlined by Zhe-Xi Luo in 2011, and clarified in 2016.[4],[5] In a review of all the data – including the new revelations from China – he suggested a radical interpretation that drew on nature's unfailing tendency to return to the same solution to a problem.

Luo suggested that the DMME had evolved independently multiple times* in different lineages of mammals. The line including monotremes split off and developed a DMME on their own later on. Meanwhile the common ancestor of therian mammals evolved a modern ear, which was the precursor to the DMME in living mammal groups. Multituberculates, therefore, must have evolved their own version of the DMME independently. Although it wasn't a perfect theory, it was much more likely that the modern ear emerged three times independently than that there had been multiple reversals, the only other way to explain the pattern. As we know already, it is not unusual for adaptations to appear in totally separate lineages to solve the same evolutionary problems.

The gain or loss of the same trait in separate lineages like this is also known as homoplasy. It remains the most parsimonious, the simplest, explanation for what we see in the fossil record of mammal ears.

---

* At first it seemed it had happened five times, but we now realise that some animals that we thought had a DMME – such as tiny *Hadrocodium*, who we met in Chapter 7 – actually had the ancestral condition.

Enhanced hearing was such a superb adaptation for survival that it developed not once, but at least three times.

Mammal ears are a Mesozoic invention not just on the inside, but also on the outside. It is likely that until they had the middle and inner ear architecture to detect high-frequency sound, the mammal lineage (and predecessors) didn't have ear flaps on the outside of their heads either. This is because pinnae are uniquely useful for determining the direction of higher frequency sounds.

In other tetrapods, sound travels through the head between the ears. This helps an animal figure out which direction it came from because the brain interprets the pressure difference in the vibrations. In amphibians and reptiles, sound travels to the opposite ear via the bones around the mouth, whereas birds have an interaural canal. But mammals are unique because each ear is virtually isolated from the other, in the sealed recording studio of the petrosal bone. This poses a problem for localising where a noise originated.

Once again, it appears that miniaturisation may have had a key role to play in solving the problem. In the 1960s researchers noticed that smaller mammals tended to hear higher frequencies. Upon closer examination, they realised it wasn't the overall body size that was important, but the distance between the ears.[6] The smaller an animal, the less time it took for sound to be detected by each ear. Mammals not only use this time difference to work out sound direction, but they also use the intensity of the sound at each ear. It turned out that when the distance between the ears was extremely small, higher frequencies were necessary for the intensity of sound to give meaningful information on direction.

This suggests that being tiny and squeaky may have been a package deal. The isolation of hearing in mammal ears meant they required more sensitive, high-frequency hearing to make sense of their world. This was the DMME. Pinnae alter the spectrum of high-frequency sound, enhancing the ability to work out if a noise is coming from in front or behind. But they don't work at lower frequencies. This means that until mammals could hear above a certain range, they had no need for ear flaps to direct sound.

The first fossil pinnae come from the Early Cretaceous of Spain, from an animal named *Spinolestes,* who we'll meet in the next chapter.

It had rounded, mouse-like ear-flaps, captured in one of the few examples of soft-tissue preservation in the fossil record outside of China. It appears these iconic flaps of skin and cartilage, which we associate with modern groups of mammals, are a relatively recent innovation.*

Being better at detecting sound would have provided a huge advantage to early mammals, helping them avoid predators, find food and communicate with one another in new ways for mating, competing or caregiving. These changes were undoubtedly tied to increased brain size. Not only did more grey matter help them process information from the senses, but they could also compute increasingly complex behaviours emerging through the Mesozoic.

We have Mesozoic mammals to thank for Mozart's arias – or Ariana Grande, whatever floats your auditory boat. Equipped with their new hearing apparatus, they had all the tools necessary to flourish and diversify in the age of dinosaurs.

The incredible range of mammal groups that launched themselves in the Jurassic built upon the revolutionary foundations of their small-bodied predecessors. Natural selection didn't skip over them just because it had dinosaurs to play with. As in modern ecosystems, Mesozoic mammals expanded into multiple habitats and their bodies changed accordingly. The shape of their teeth would prove a key contributor to this success. For docodontans, teeth opened the door to developing the first specialisations for dedicated swimming, climbing and digging.

As the Jurassic drew to a close, the startling docodontans and parachuting haramiyidans were winding down. Other mammal groups were emerging in their place, and went on to flourish for the next 80 million years of the Cretaceous. Among them were the earliest members of the clade Mammalia. This included the ancestors of the deep and hardy monotreme branch. They were one of the first groups

---

* Some modern mammals have no pinnae. They are usually aquatic (such as cetaceans and seals), or don't localise sound, often because they live underground (such as moles and pocket gophers).

to diverge from the pack, living alongside the many members of the Mesozoic mammalian tree – most of whom died out along with the non-bird dinosaurs.

For the mammals thriving in this lush landscape, their time in the Mesozoic sun was bright and prolific. And as the Cretaceous dawned, their world was ready to blossom.

CHAPTER TEN

Time of Revolt

Which one amongst them did a promise hold
That Mammal life might issue from its fold?
...
And may we not to that old stock assign
The founding of the mighty mammal line?

Henry Knipe, *Nebula to Man*

The Atlantic Ocean washed relentlessly over the shingle coast of Britain. Just beyond the horizon, German U-boats* waited for British Allied shipping targets. It was September 1939. In these early months of the Second World War, Germany's tactic of bombing supply ships making the transatlantic crossing was proving devastatingly effective at cutting off Great Britain from her allies.

A patrol of soldiers walked the shore, diligently guarding their coastline from the new threat. They saw a man walking along the base of the cliffs. He was carrying a map and scanning the landscape. Unnerved, the soldiers descended to the beach to confront him. Their alarm grew further when they heard his accent: *German*. Despite his protests of innocence, they immediately arrested him on suspicion of being a spy. Why else would he be wandering their vulnerable beaches clutching a map?

The man was Walter Georg Kühne, 'legendary German searcher for Mesozoic Mammals'.[1] At the time of his arrest, he was in his late twenties, and unknown to the concerned soldiers, a well-regarded fossil collector.

Kühne had arrived in Great Britain from his native Germany the previous year along with his wife, Charlotte. His career in Germany at the University of Halle had been cut short by the tense political atmosphere of the 1930s. He was expelled from the University and

---

* *Unterseebooten* (submarines).

accused of communist sympathies, spending nine months in jail.[2] Afterwards, he and Charlotte funded themselves by collecting fossils and selling them to museums. As things became more dangerous politically they decided to take refuge in Great Britain, and continued to make a living from fossil collecting.

The Kühnes seemed to have a talent for locating the tiny, rare bones of the world's first mammals. They'd already had huge success screen-washing material from fissure fills at Holwell Quarry and several other Triassic and Jurassic sites in England and Wales. The Triassic mammal *Morganucodon* was found and named by him. One of his enduring legacies was the development of screen-washing processes for recovering Mesozoic mammal fossils.[3] 'I know where to discover early mammals,'[4] Kühne affirmed to Francis Rex Parrington, vertebrate palaeontologist at the University of Cambridge. Parrington himself was an expert in the evolution of mammals, and delighted by the fossil teeth Kühne brought to sell to the Museum. He offered £5 (over £300 in today's money) for every early mammal tooth found.

That September, just after Great Britain and France declared war on Germany, Kühne had gone in search of new fissure fills in the eroded Atlantic cliffs with the aid of a geological map and hammer. That was where the soldiers found him. He would spend the rest of the war in an internment camp on the Isle of Man.

Thankfully, Kühne had made many friends in the scientific community in Britain. Colleagues from the Natural History Museum and multiple universities in London petitioned the government to allow him to continue working with them during the war, and even to visit the museum in London despite his internment. Kühne used his time to work on his fossils, and put together one of his most important contributions to cynodont palaeontology: a monograph on a close sister to mammals, named *Oligokyphus*.[5]

Kühne's monograph became a touchstone for palaeontologists. He provided detailed figures and descriptions of the anatomy of this creature, pieced together from over 2,000 separate isolated bones. Because *Oligokyphus* is from a group on the periphery of the mammal family, its body plan provides vital clues to help us understand the common ancestor of mammals and non-mammalian cynodonts. Even today this animal is used in hundreds of phylogenetic analyses as a comparison point, an outgroup, to the families of mammaliaforms and mammals. *Oligokyphus* is the gold standard of *almost*-mammalness.

Despite this, *Oligokyphus* is forgotten by most palaeontologists and the public. Its kin are an exceptional example of specialist adaptation in cynodonts, but nowadays they are mainly remembered for their superficial resemblance to the modern world's most common mammals, the rodents. *Oligokyphus*'s diet and skull shape, however, have become the most successful in the whole of mammal history.

◇ ◇ ◇

Oligokyphus belongs to a group called the tritylodontids. They are among the branches of cynodonts thought to be the closest relations to mammals, and their fossils are some of the most prolific in the world. Kühne found them in Britain, but they are known on almost every continent, from North America and Asia to Africa and even Antarctica. They ranged from modest, weasel-like creatures to those that could take on a honey badger and win. This group existed for an incredible length of time, spanning the end of the Triassic until at least the Early Cretaceous. With 80 million years under their belt, they are some of the most persistent Mesozoic cynodonts. The details of their skeletons are pivotal for understanding early mammal origins and relationships. This is why Kühne's monograph on *Oligokyphus* was worth waiting for, and remains useful to this day.

Despite all this, tritylodontids are the underdogs of cynodont palaeontology. This is strange, because as well as being successful, they possessed specialisations for a lifestyle unexplored by their earliest mammal contemporaries: herbivory. By the Early Jurassic the first tiny mammaliaforms were dining out on insects. All of the non-mammalian cynodonts had turned to dust, except for the tritylodontids, which flourished. Natural selection had moulded them into premier leaf-grinding machines.

Of course, there had been synapsid herbivores before. The mighty gut-busters of the Late Carboniferous had figured that niche out first, and Permian hulks honed it, fermenting their endless leafy feasts in guts riddled with symbiotic bacteria. Dicynodonts kept the torches burning through the Triassic. But those bulky predecessors were gone. In the world of crocs and dinosaurs, mammals had microscoped themselves into the understorey, and the tritylodontids were left to eat the scenery.

It is thought that this specialisation for plant-eating was an important reason for their success where all other non-mammalian cynodonts died out. Almost all of the early mammaliaforms sharpened their cusps and became carapace busters. Their teeth were scissors and squashers, with a once-in-a-lifetime replacement to ensure they snipped and smooshed to maximum effectiveness.

Tritylodontids took a different route. Their postcanine teeth were shaped like parallel mountain chains, aligned to run the length of the jaw. The peaks of the upper teeth sat in the valleys of the lowers. As they chewed, the jaw slid from front to back, grinding greenery into the enamel troughs. Their mouths were also continuous escalators of tooth replacement. New teeth arose from the back while the old teeth dropped out at the front, endlessly replenishing the mountain chain. This constant cycle of postcanine replacement compensated for the continual wearing down of the teeth, caused by a heavily fibrous diet.

A conveyor-belt replacement pattern has also evolved in some modern mammal groups, although the mechanism is somewhat different. Elephants, for example, replace their enormous molars from the back of the jaw, with the new ones ousting the old. This happens up to six times in their long lifetime – if they outlive their sixth set, they usually die of starvation. Sirenians, better known as manatees or sea cows, do the same. They are closely related to elephants, but unlike them appear to have an unlimited supply of molars, bringing them in one at a time until they have a full dentition.[*] Some kangaroos do something similar. Continuous replacement is not the norm for mammals today, but in each of these cases animals

---

[*] The elephant and sirenian lineages split around 50 million years ago. Intriguingly, it appears that elephants may come from a semi-aquatic ancestor. The teeth of a 35-million-year-old fossil elephant relative called *Moeritherium* indicate this animal fed on aquatic plants. Scientists know this thanks to carbon isotopes in the teeth, which match their food. Although not a direct ancestor of modern elephants, there is a chance it represents the way of life for many early members of the group, which would mean the lineage went from being terrestrial, to aquatic, then fully terrestrial again. Come on Dumbo, make up your mind.

have developed this modification in response to the heavy wear caused by herbivory.*

There is more than one way to beat the daily grind. For the world's most successful lineage of mammals the solution comes not from the back teeth, but the front. From an evolutionary point of view you could consider this group to be the ultimate pinnacle of all synapsids. It's an assembly of the most radical innovations that natural selection can offer: combining small size with the burning fires of endothermy, fast reproduction that allows their numbers to explode, a body plan that makes it possible to move in almost any habitat, and ever-growing incisors to gnaw their way through everything from the toughest seeds, to entire tree-trunks. I am, of course, referring to the rodents.

Almost half of the mammal species on Earth today are rodents. From nests of mice to scurries of squirrels, porcupines, beavers, jerboas, guinea pigs and capybara, rodents are everywhere.[†] Their name comes from the Latin *rodere*, to gnaw. We recognise them by their prominent front teeth, elongated and thick like two woody stems. Inside the skull these teeth are rootless, emerging incessantly from the jaw as fast as they can be ground down. They are perfectly adapted for feeding on plant materials, particularly tough ones that need a good gnawing. The agouti of South America for example (*Dasyprocta*) is one of the only animals capable of getting into the almost impenetrable Brazil nut. Beavers (*Castor*) can fell trees, chewing through their trunks and hauling them off to build their homes and feast on their sweet upper leaves and bark.

The rodent skull is instantly recognisable. Those big front teeth (sometimes stained orange with iron pigments, which strengthens them further) sit in front of a toothless gap. Behind are a row of flattened, grinding molar teeth. The skull has the overall appearance of an ornate pincer, the whole structure seeming to focus on maximising the incisor bite. That huge toothless gap behind the front

---

*Another way to solve the problem is to develop tall-crowned teeth to withstand a lifetime of wear.

[†] Except Antarctica, but with global warming it's just a matter of time.

teeth is called a diastema, and marks the 60-million-year-old spot where the canines would have grown in their ancestors.

Herbivores often lose their canines and develop a diastema: from the ancient therapsids to modern grazers, it's a tell-tale mark of the mulch-muncher. Rabbits and hares have it too, and are often confused with rodents despite belonging to a different taxonomic group (the lagomorphs). The confusion is understandable, as both have similar skull shapes and ever-growing front teeth. Superficial resemblances like this are a recurring theme in nature. They often confuse our initial attempts to classify animals. The similarity of shapes comes from evolutionary convergence: nature handing out her patented solutions for succeeding at the same lifestyle.* In this case, dedicated gnawing and grinding of plant matter.

Casting our sights back to the Mesozoic, we see the same 'rodent-like' skull in tritylodontids – but with a splash of elephant. Their incisors were enlarged with a diastema behind them, forming the distinctive pincer-shape. Their grinding teeth at the back were replaced in a conveyor, seemingly throughout their lives. In many places, this group is known only from their shed teeth, which litter fossil faunas around the world – including the Isle of Skye in Scotland[6] – like discarded cigarette butts.

Conveyor-belt teeth in tritylodontids were a modification of the pre-existing continual tooth-replacement of Triassic cynodonts. For the modern groups that replace teeth in this way, like manatees, they have returned to a pattern akin to continual replacement by abandoning the diphyodonty of their Mesozoic ancestors.

This feature tells us not only that tritylodontids were highly specialised for herbivory, but suggests they may not have fed their young on milk. There is no single wholesale tooth replacement in youth, and no evidence for a single weaning event either. Instead, baby tritylodontids had teeth just like their parents – only tinier. They were born to browse. We know, thanks to a discovery published in 2018 by palaeontologists Eva Hoffman and Timothy

---

* You may notice I mention convergent evolution repeatedly throughout this book. It's no accident; by constantly bringing it up like this, I hope I give you a taste of its prevalence in evolutionary history.

Rowe, that tritylodontids also had huge litters of babies: a fossil of *Kayentatherium* from Arizona in the United States was found with at least 38 babies nestled underneath it.[7] This is well outside the largest litters of most mammals,* and suggests that it was after the Late Triassic that litter size in mammals reduced in numbers. It is likely that big litters were another legacy from the ancient amniote past.

Tritylodontids are surprising because they endured alongside mammals for so long when all other non-mammalian cynodonts died out. But like all of us in the grand story of life on Earth, their days were numbered. Overlooked and under-studied, tritylodontids are remembered as foreshadows of the groups that replaced them. They were knocked from their position as the top herbivores by a group that would become the most diverse plant-eating mammals of the Cretaceous – and beyond.

In the Middle Jurassic to Early Cretaceous a branch of 'true' mammals swept across the world. They too had rodent-like pincer skulls and grinding teeth. Although only distant cousins, they exploited the same adaptations as the tritylodontids who preceded them. Their body plan would carry them all the way out of the Mesozoic and through the asteroid impact, to compete with their brethren for the top slots in the 'age of mammals'. To this day, they are also the record holders for clade longevity.

And they owe it all to flower power.

◌ ◊ ◌

High over the desert of Southern California, one of the world's most expensive cameras floats through space. With a pixel size of 15 metres (49ft) it's hardly going to take a flattering photo of you or I, but every 99 minutes it circles the Earth from over 700 kilometres (438 miles) away, capturing over 720 images a day.[†] It is the satellite Landsat 8, and it takes our planet's ultimate long-range selfie.

---

* The closest is the tenrec, a small mammal from Madagascar, some species of which habitually have around 32 babies. This is unusual among mammals though, and most litters are limited by the number of nipples available for suckling. Marsupials often have more babies than nipples – and not all of the young survive.
† Only slightly more than some people on social media.

At the end of March 2019, the Earth blushed under all the attention. Gazing down at the normally pale complexion of the Colorado Desert, a flush was rising. Flowers were blooming across vast swathes of arid landscape. Apricot-petalled poppies, lilac verbena and the rich butter of dandelions spread over the hillsides and coated the valleys. The desert had sprung to life.

It is called a superbloom, a natural phenomenon usually only seen once per decade. It happens when the weather conditions are just right, with plenty of winter moisture. For the previous seven years there had been very low rainfall, and for two years drought across Southern California had killed off most of the arid scrubland and invasive dry grasses that dominate the landscape. An especially wet winter kissed awake millions of sleeping flower seeds. As spring arrived, the once-parched hills erupted in a blaze of colour. The scale of superblooms is so immense that they can be seen from space.

On the ground the floral spectacle attracted swarms. Not just pollinators, but an equally prolific pest: Instagrammers. Social media exploded with news of the biggest #SuperBloom in 30 years. Over 50,000 tourists crowded into the blooming hotspot of Walker Canyon, just north of the small town of Lake Elsinore.[8] Influencers queued for an hour and a half to get in, and vied for spaces to pose. They snapped themselves performing yoga amid the mosaic of blossoms. They lay down, crushing the petals as they pouted up into their camera phones. They wanted not just to capture it, but capture themselves *in* it. Nature, in this new age, is nothing more than the glorious backdrop for our next duckface.

There would be no humankind, let alone superbloom selfies, without a key event in Earth's evolutionary history. Around 120 million years ago, a group of plants emerged that utterly transformed both the literal and the ecological landscape. It is certain to say that without them we wouldn't be here. I am, of course, talking about flowers.

Look out of the nearest window. I'll stake my next month's income (not a high wager, so don't get excited) that the first plant you see is an angiosperm: a flowering plant. The name comes from the Greek *angeíon*, meaning case or bottle, and *sperma*, a seed – in other words, a fruiting plant. Angiosperms are such a massive component of our world today that we are almost blind to them. We poison them, pluck them, watch them die in vases and eat them. We walk on them, sit on them and sit under them. Meanwhile they make the very air we breathe.

Believe it or not, there were no flowers until at least 130 million years ago. For the first 350 million years after plants first made it onto land they had been blossomless. As we saw in earlier chapters, the 'floral' component of ecosystems in the past was first dominated by lichens, clubmosses, quillworts, ferns and then gymnosperms. Members of these groups are still alive and well today, and of the latter, conifers remain a core component of forests the world over. The gymnosperms produce seeds and pollen, which led to some mutual relationships with pollinating beetles and other insects. But for the majority of these plants, they did a good job at reproducing by means of wind and water alone.

The exact origin of the first angiosperms is unclear. The oldest fossil pollen from an angiosperm comes from the Early Cretaceous, but the oldest fossil of an actual flower is disputed. In 2018 a team of international palaeobotanists announced that they had found it in the Early Jurassic Xiangshan Formation in China.[9] This comes closer to the estimates of flowering plant origins based on genetics, which suggest that they could have diverged from among the gymnosperms as early as the Triassic.[10,11] One theory is that the genome of the ancestral angiosperm duplicated itself – a mutation that has been suggested as a major driving force in the evolution of many plant and animal groups. Although some pollen has been found in Middle Triassic rocks of Switzerland that might have belonged to a flowering plant,[12] there is no solid evidence to support such an ancient origin. Other researchers have dismissed the Early Jurassic fossil as a misidentified conifer.

Scientists believe that the first angiosperms, whenever exactly they appeared, were adapted to living in wet environments, deep in the dark understorey of forests. They probably had low population sizes, and were geographically restricted. These first angiosperms quietly paddled along in the background throughout the Jurassic, a small component of a gymnosperm-dominated world.

In the Cretaceous, however, flowers came of age. Their superblooming in the fossil record is so instantaneous that it confounded the earliest evolutionary scientists. Darwin had asserted that evolution could 'produce no great or sudden modification; it can act only by very short and slow steps'.[13] Yet there they were, appearing fully wreathed out of thin air in the Middle Cretaceous, as though dropped from the sky. He called it 'an abominable mystery', and corresponded with botanists around the world trying to solve it.

As our understanding of evolution has refined, the abrupt appearance of flowering plants is no longer a puzzle. Darwin was only half right – natural selection may be slow and incremental, but it can also race and explode given the right circumstances. The low resolution of the fossil record means it can take a great deal of searching to find 'transitional' fossils to illuminate such a short space of geological time, but sooner or later they usually turn up.

There are three fossils that address Darwin's mystery. One is from Portugal and belongs to a water lily (Nymphaeales).[14] Another, called *Montsechia*, comes from neighbouring Spain, and resembles pondweed.[15] The third is a relative of the buttercup (Ranunculales), called *Leefructus*, and it was found in Liaoning Province in China.[16] All three are from the Early Cretaceous, around 125 million years ago. Plants with flowers were clearly already present at that time, including members of some modern groups. They didn't appear from nowhere after all.[*]

It was then that flowers ignited their global revolution. Between 125 and 80 million years ago, from their damp understorey habitats the angiosperms suddenly emerged in one of the most transformative explosive radiations in the history of life.

This event is part of what has been dubbed the Cretaceous Terrestrial Revolution, and it was utterly wholesale. As the basis for habitats and food chains, changes in plant communities alter entire ecosystems. Insects, mammals and reptiles in the Cretaceous all adapted and diversified as angiosperms produced new and abundant food sources like nectar and fruit. The Earth's heart was suddenly throbbing with perfumed juices.

The first pollinating bees appeared in the Early Cretaceous, evolving from a predatory wasp-like ancestor. Moths and ants diversified, as

---

[*] Angiosperms are split into two groups, the monocots and dicots, and it is worth pointing out that it is the diversification of the dicots that Darwin was specifically referring to as a mystery, rather than all angiosperms.

did grasshoppers and gall wasps. All of these provided food for new predatory insects, and for those that fed upon them all. New lineages of lizards and snakes, as well as their larger cousins, the crocodiles, crept and slithered into existence and proliferated.

Dinosaurs were also doing well, although it is thought that their continued evolution into new forms was probably not driven directly by the floral revolution going on around them. By the end of the Cretaceous, herbivorous groups that are synonymous with the whole idea of dinosaurs – such as the neoceratopsians (*Estemmenosuchus**-wannabes, like the horned *Triceratops*) and club-tailed ankylosaurs – were munching on the low growth. Above them browsed the head-butting pachycephalosaurs, bipedal herbivores with domed bony skulls fringed with horns like a hellish monk's tonsure. Hadrosaurids were pulling the first duck-faces amid the Late Cretaceous superblooms. Instagrammers would have been a tasty snack for the more notorious wildlife of the time, for this was when the classic tyrannosaurs and spinosaurs were afoot, those predators that obsess Hollywood and small children in equal measure.

More interesting, however, are the members of a group of dinosaurs that still exist today. The ancestors of birds, called avialans, first appear in the same Late Jurassic sites that hold such a stunning diversity of swimming, gliding and burrowing mammals. By the Early Cretaceous, fossils like *Jeholornis* and *Sapeornis* from China flew in to fill the gaps between the spindly bipedal maniraptoran dinosaurs of the Jurassic, and animals like the Late Cretaceous 'wonderchicken' (an early bird close to the common ancestor of chickens and ducks).[17]

Mammals were also swept up as the Cretaceous budded. The revolution of plants had huge impacts on herbivores in particular.

The last known tritylodontid fossils come from a Cretaceous site in central Japan. Around 130 million years ago a system of rivers braided their way across what is now the Shiramine District, 400 kilometres (250 miles) west of Tokyo. In a side-channel the water washed the bones of animals to their final resting place: frogs, turtles and lizards, even the occasional dinosaur. Among the debris lay little cuboid teeth with rows of cusps and valleys. These were all that remained of

---

* We met this bizarre therapsid in Chapter 5.

Montirictus kuwajimaensis, the last tritylodontid.[*] Its teeth are found beside those of its replacements, inheritors of the herbivorous cynodont niche: the multituberculate mammals.

Multituberculates became the most prolific, numerous mammals of the latter part of the Mesozoic. To understand their place in the changing world of the Cretaceous, we must return to the heart of a flowering desert, and the woman who would shape our knowledge of Mesozoic mammals and carry it into the 21st Century.

 o ◊ ✿

Revolutions shape people as well as nature. While Walter Kühne was studying *Oligokyphus* under the watchful eye of the government in wartime Britain, a Polish teenager was joining the largest European underground resistance against Nazi occupation. Her fire and steel were forged fighting Nazis, and later her determination grew by resistance to the insular muffling of the communist regime. She was the first woman to lead palaeontological expeditions into the desert regions of Mongolia. No mere female Indiana Jones, her intellectual rigour and collaborative approach transformed our understanding of mammal anatomy and evolution around the world. She was one of the world's great scientists and the doyenne of Mesozoic mammal palaeontology, and after a career spanning six decades, in her eighties she literally wrote the book on Mesozoic mammals.

Her name was Zofia Kielan-Jaworowska.

Zofia Kielan was born in 1925 in the small town of Sokołów Podlaski, halfway between Warsaw and what is now the Belarus border. The Kielans claimed mixed European ancestry. Scandinavian blood on her father's side had been traced back to Swedes who fled to south-east Poland in the mid-1600s. Her mother was descended from minor Scottish nobility, or *duniwassal*, from the Gàidhlig *duine* (man) and *uasal* (noble). Zofia Kielan was a citizen of everywhere and nowhere, a child moulded by conflict with the power to help rebuild the world.

---

[*] As this book went to press, another tritylodontid from the Early Cretaceous was named by my colleague, Fanguan Mao, and her co-authors. It is a gorgeous skeleton from China, *Fossiomanus sinensis*, and it joins *Montirictus* as one of the last of its kind.

Kielan's father Franciszek had risen to a stable position as chief warrant officer working in Russia. The October Revolution of 1917 prompted him to return to Poland in 1918, where he met Maria Osińska while he was still serving in the First World War. After the war the couple had two daughters, Krystyna and Zofia, and they all moved to Warsaw when the girls were still babies. With his new job and Maria working in an office, between them they brought in a good income that allowed the family to buy a home.

Kielan's strong character and academic gifts were clear from a young age. In photographs she has a serious air, lips pressed together in determination. Her treacle-eyes challenge you defiantly. One Christmas as a child, when she disapproved of a large family gathering taking place at their aunt's home, Kielan insisted on sitting as far away from everyone as possible, maths textbook on her lap, obstinately completing classroom assignments while the rest of the family celebrated.

But this diligence paid off. At the age of 11 she was accepted to a prestigious school in the city. She spent hours poring over textbooks on zoology and palaeontology in the Zoological Museum library. The theory of evolution lit a fire in her mind, and she determined to make it her career. The family were on course to live comfortably and prosper.

The year 1939 would transform the Kielans' lives, as it did a whole generation. Zofia Kielan was 14 years old when Germany invaded Poland and began its air-assault on her city. In September of that year Nazi forces entered Warsaw and Poland came under German and Soviet Russian rule.

It was in this war-ravaged landscape that Kielan's formidable personality was moulded. Later, she remembered the year the war began as significant not only because of these world-changing events, but because it was the year she left the Roman Catholic faith, concluding it was 'illogical and full of contradictions'.[18] Her parents accepted her refusal to attend church easily, being non-practising themselves. It was still a bold declaration for a young woman in a country that remains predominantly Roman Catholic to this day, but Kielan was never one to shy away. She remained a firm atheist for the rest of her life.

At 15, she trained as a medic in the underground paramilitary group the Grey Ranks. Secondary education was banned for non-Germans by the Nazis and punishable by death. Yet the Kielan girls continued learning thanks to their headmistress, who arranged clandestine classes under the guise of a permit to teach German language and gardening – the only subjects (along with knitting) permitted to be taught to girls.

For the next five years the Kielan family continued to defy the Third Reich. They sheltered two of their daughter's young Jewish schoolmates, helping one escape to safety while the second, Jana Prot, remained in their home, her background undetected.* When the Warsaw Uprising finally approached, Kielan and Jana were assigned to treat some of the 6,000 injured Polish soldiers who fought to drive the occupying force from the city. Despite the chaos around her, Kielan often travelled with a palaeontology textbook in her backpack, which she read from in quiet moments. She still hoped to one day pursue her dream.

The uprising failed. Kielan and Jana were taken to a camp, one step away from transport to a forced labour camp. There, they found Kielan's mother, and the three of them narrowly escaped to safety in the town of Skierniewice. But the destruction of Warsaw at the hands of the Nazis was extensive. As many as 200,000 civilians died, many murdered in mass executions. 'The city must completely disappear from the surface of the Earth,' Heinrich Himmler had decreed, 'no stone can remain standing. Every building must be razed to its foundation.'[19] In their genocidal master plan, Warsaw would become little more than a transit hub, its people relocated or killed.

By the end of the war, four-fifths of the city was gone; some in fighting, but much of it through systematic, planned demolition. Historic monuments, libraries, schools, universities – nothing was spared. When the war finally ended, the Soviet army entered the tattered remains of Warsaw, and the city and country remained part of Russia's communist Eastern Bloc for the next 54 years.

At the war's end Kielan walked the 71 kilometres (44 miles) back to Warsaw from Skierniewice. She found the family apartment destroyed.

---

* In 1991, the Yad Vashem Institute (dedicated to Holocaust research) awarded Franciszek and Maria Kielan and their daughters the title 'Righteous among the Nations'. This award is presented to the rescuers of Jewish people in recognition of their actions. It was an award Kielan deeply cherished.

Dragging her miraculously undamaged bicycle from the rubble, she rode to the Zoological Museum. There she joined the homeless museum employees who had set up temporary shelter among the remains of the drawers and exhibits. Together they dug out the collections. Piece by piece, they began to rebuild and restore the museum.

It wasn't long before Kielan became an employed assistant there. The museum's library had somehow escaped destruction, allowing Kielan access to a wealth of stored knowledge. She continued her studies and completed her doctoral degree.

Surprisingly, her initial research was not on the early mammals for which she would later become famous, but on trilobites and ancient worms. Having completed her PhD, Kielan-Jaworowska – for she had married her best friend, medical student Zbigniew Jaworowski, in 1958 – wanted to return to the study of the vertebrates that had captured her heart as a teenager. But her supervisor had other plans. He suggested she focus on polychaetes.

Polychaetes, also called bristle worms, are often known in the fossil record just from their biting jaws – the only hard parts in an otherwise soft and squishy body. Kielan-Jaworowska used acid to remove some of the first complete fossil polychaete jaws from rock, allowing her to reconstruct ancient bristle worm mouths in exceptional detail. Her eye for minute structures and patience for difficult work laid the foundation for research on the mammal fossils she would discover in years to come.

By the 1960s Kielan-Jaworowska had risen to the position of director of the Institute of Paleobiology in Warsaw, part of the Polish Academy of Sciences, Polska Akademia Nauk (PAN). Her predecessor, an open and internationally minded man, had just completed negotiations to begin joint Polish-Mongolian palaeontological expeditions to search for fossils in the Gobi Desert. Kielan-Jaworowska knew of the region from the accounts of her childhood hero, Roy Chapman Andrews.* He had journeyed there in the 1920s as part of a

---

* Andrews is thought to have been the inspiration for the character of Indiana Jones, from the namesake movies. This has never been confirmed, and another contemporary figure, the British explorer Percy Harrison Fawcett, has also been suggested. Both present the stereotypical male adventurer, one that archaeology and palaeontology alike are trying to shake off as we thankfully move away from the macho plundering of the past.

team from the American Museum of Natural History, discovering the fossils of dinosaurs and their eggs. But for Kielan-Jaworowska it was other treasures that captured her imagination.

'I still distinctly remember the day I first heard about paleontological expeditions to the Gobi Desert,' she wrote later in her book, *Hunting for Dinosaurs*. In 1946 'in a small room in Professor Kozłowski's own flat on Wilcza Street ... We arrived one morning to find drawn on the blackboard the skulls of two mammals with strange names *Deltatherium* and *Zalambdalestes*. These were the earliest placental mammals known ... It was then that I learned for the first time that the Gobi was a veritable Eldorado [sic] for palaeontologists.'

As a girl Kielan-Jaworowska had been gripped by accounts of the Gobi expeditions. Her love of adventure and fascination with travel grew as she hunted articles and pored over Andrew's book of his global exploits, *To the Ends of the Earth*. 'Not in my wildest dreams did I expect that I, too, would someday go there, and that only 16 years later I would be organising a Polish-Mongolian paleontological expedition.'

In 1962 Kielan-Jaworowska was given six months to prepare an initial exploratory team. For most Western scientists at the time it was almost unthinkable that such a pursuit could be led by a woman. Similar campaigns from Europe, the United States and South Africa were all headed by men. But Kielan-Jaworowska was one of several pioneering female scientists in Eastern Europe who trail-blazed in all aspects of their disciplines.

The 20-page equipment list she drew up included a three-month supply of fuel for their vehicles, food, all the necessary camping and cooking equipment, medical supplies and clothing suitable for scorching days and freezing nights. It was packed in crates and sent via the Trans-Siberian railway to the Mongolian capital, Ulaanbaatar, to await their arrival.

The team needed to prepare academically as well, but there were no dinosaur or Mesozoic mammal fossils held in any collection in Poland at the time. Most of the books and scientific papers on them had been destroyed in the war. So Kielan-Jaworowska had key documents imported on microfilm. The team held seminars together, discussing the geology and palaeontology of the Gobi as they waited out the cold Polish winter. They learned the stratigraphy and drew up lists of all the dinosaurs and mammals from similar-aged sites around the world.

In May 1963 Kielan-Jaworowska's first Polish expedition boarded their flight from Warsaw to Ulaanbaatar without her. They were on a reconnaissance mission. With their collaborators in Mongolia they scouted out promising sites and learned hard lessons about which vehicles could reliably traverse the desert. They designed tents that were suitable for the baking midday sun and ceaseless winds. They returned four months later with initial, surface-collected mammal fossils from the Palaeocene (56–66 million years ago), and a few dinosaur eggs.

Armed with the reconnaissance team's information, Kielan-Jaworowska prepared for a second expedition to Mongolia in 1964 – and many more would follow. Finally her time had come: she was ready to explore the arid, Cretaceous heartland of Asia.

◦ ◊ ◖

The Cretaceous takes its name from the Latin *creta*, meaning chalk, but it is often abbreviated to K in geology, from the German for chalk, *Kreide*. The end-Cretaceous mass extinction event is called the K–Pg (Cretaceous–Palaeogene), and the Cretaceous Terrestrial Revolution is the KTR. Deep deposits of this soft white chalk outline the body of the long-dead era across Western Europe. Chalk is formed from calcium carbonate in the shells of marine invertebrates, telling us that Europe was submerged under a rich shallow sea at this time.

Europe was not alone in the flood. There was a widespread inundation in the Cretaceous, as the last remnants of the supercontinent of Pangaea finally shattered. Not only did the oceans we recognise today take shape – North and South Atlantic, Pacific, Arctic – but the continents were also assuming positions we are familiar with. Despite this, we might still fail to pick our Earth from a line-up. India continued to dally off the coast of southern Africa, hugging Madagascar to its chest. In the furthest southern latitudes, Australia and Antarctica were also having a hard time letting go.

Thanks to the highest sea levels since the Carboniferous, the continents of the Cretaceous had radically different coastlines. Northernmost Africa, the Middle East and almost all of Europe were underwater. North America was cut in half, north to south, by the Western Interior Seaway. At times this joined up with the Hudson and Labrador seaways, slicing the continent into three. In South

America, most of the north-west was also marine, with seas invading Venezuela, Columbia, Ecuador and Peru.

The world's most famous mountain ranges didn't exist. The Alps were only just being birthed from the rich remnants of the Tethys Sea. They folded and lifted as Africa and Europe drew closer. Along the eastern edge of the Pacific Ocean the Rockies in the north and the Andes in the south were mere swellings left over from orogenies* past.

But the ground was rising again. In the Cretaceous the tectonic plates along the eastern Pacific rim started to plunge under one another, causing the crust to lift, a process that continues to this day. The Andes is now the longest continental mountain chain in the world at around 7,000 kilometres (4,300 miles). The Rockies have been eroded by subsequent ice ages, but at their zenith would have topped the gasping heights of today's Tibetan Plateau. The Himalayas, now the benchmark of all things mountainous, were nowhere to be seen. They are geological babies, only appearing in the last 50 million years, and still on the up.

All of these mountains play crucial roles in shaping our climate. They alter air currents, steal rain and funnel weather systems. They also lock up water in the form of glacial ice, as does the South Pole. Although some glaciation is thought to have been present on Antarctica during the Cretaceous, for the majority of the 80-million-year span of this time period the Earth saw little snow or ice. This added to the raised global sea level, turning our planet into a great water-world. Ocean currents could race a gauntlet around the equator from east to west almost unhindered.

These geographical changes combined to raise global temperatures. The Earth emerged from a cooler period at the end of the Jurassic and it steadily heated until even Antarctica was green. With little difference in temperature from the equator to the poles, ocean upwelling slowed and parts of the sea stagnated. The deepest oceans were as much as 20° Celsius warmer than today, leaving deposits of anoxic shales in the rock record.

Despite this, the plentiful marine habitats formed by raised sea levels hosted a similar revolution below the surface to the one that was taking place above. Since the beginning of the Mesozoic, a shift had

---

* The plural of orogeny, the convergence of pieces of the Earth's crust.

been occurring in our oceans. In the early days marine diversity was underpinned by sedentary life on the sea floor. Creatures had done well building exoskeletons and shells and hunkering down. But as the Triassic dawned, these organisms were being picked off by new breeds of marine predator. Shell-crushers, specially adapted to break hard bodies between pebble-like teeth, were feasting their way through the food chain. Those animals that couldn't adapt to the new threat were wiped out. Only the more mobile or well-hidden survived.

At the other end of the food chain, the iconic marine reptiles – not dinosaurs, but a separate radiation that fully returned to underwater living – emerged and flourished. Ichthyosaurs, plesiosaurs, pliosaurs and mosasaurs all filled the oceans of the Mesozoic. They were among the first tetrapods to achieve this feat of aquatic adaptation, which would later be echoed in the lineage of mammals that became whales and dolphins.

Mongolia was above sea level during the Cretaceous. It was high land, but not dry land. This is in stark contrast to today's country.

The steppe of Genghis Khan, landlocked between Siberian Russia and China's Inner Mongolian Province, is now the epitome of Western notions of harsh 'remoteness'. Like Timbuktu, Mongolia is often a byword in the English language for the middle of nowhere – despite both locations being centres of great and expansive ancient empires. It remains one of the least densely populated nations in the world, with only three million inhabitants, a third of whom are nomadic or semi-nomadic herders.

Much of the country is high-altitude grassland, varying from the arid flatter southern reaches to the high mountains of the northern and western border with Russia. In summer, temperatures scratch the high 40s Celsius (over 115° Fahrenheit), whereas winter cracks a frosty Siberian whip at -30° Celsius (-22° Fahrenheit). The cruelty of this season is such that Mongolians have five different words for mass starvation of their livestock.*

---

* These words for mass starvation, or *dzud*, are *Khuiten* (cold), caused by severe cold; *Tsagaan* (white) caused by high snowfall; *Khar* (black) caused by lack of water; *Tumer* (iron) resulting from a brief melting followed by freezing temperatures, locking the grazing in a sheet of ice; and *Khavsarsan*, a combination of two or more of the other *dzud*.

Although we call it the Gobi Desert, the word gobi has a very specific meaning. Unlike a sandy desert, gobi is a grassland that is so dry that although it can support camels, it can't support marmots.* In other words, it is a fragile and sparsely vegetated, water-poor environment. The desert lands south of Mongolia's Gobi sit in the shadow of the Himalaya. Those mountains greedily drink the southern rains and parch the land to the north, from Tibet and the Taklamakan and Gobi deserts to the belly of Mongolia itself.

But Mongolia wasn't always like this. It is a landscape trapped by geological circumstance. The rocks and fossils found in the Gobi are like sepia photos on an elderly relative's mantelpiece. In them we see a lush and vivacious Mongolia, decked in her finest jewels, the life and soul of the ecological party. This was a landscape working with all that the Cretaceous had to offer. The fossils found there would shape our understanding of mammal evolution, the origins of modern mammals, and the careers of those who studied them.

◌ ◇ ◌

Kielan-Jaworowska arrived alone in Ulaanbaatar on a scorching June day in 1964. She understood why Mongolia is also known as the 'Land of the Eternal Blue Sky', with more than 250 cloud-free days every year. After a week working with scientists from the Mongolian Academy of Sciences and the University of Ulaanbaatar, the Polish palaeontologist set off south with two companions in a small field car.

A local employee of the Academy, Mr Dagwa, sat in the back seat. He spoke Polish from a year spent in Kielan-Jaworowska's home country. In the driver's seat was Batochir, another Mongolian who also spoke Russian. Dagwa and Batochir chanted Mongolian folk songs as the trio sped along the faint tracks made by previous vehicles in the sparse flat steppe.

The Tolkienian map that accompanies Kielan-Jaworowska's account of this and subsequent expeditions charts the long journey to the hot south-west of the country. For two days they rode past horse and camel herds, arriving at the town of Dalandzadgad, 550 kilometres

---

* Marmots are chubby ground-dwelling squirrels, common in grasslands and high mountains.

(340 miles) from Ulaanbaatar. From there, it was another few days' drive west into the Nemegt Valley, where her advance team was already hard at work. Gradually the green grasslands petered out and they entered the desert. When the roads disappeared, they drove along dry river beds.

Kielan-Jaworowska took to Mongolian nomadic life with ease. The scarcity of water and communal accommodation came as no surprise to someone who'd so attentively read Chapman Andrew's accounts of travelling in the region. All of their water had to be boiled before use, and to make a fire they gathered *Haloxylon*, a desert shrub that burned bright and hot. They enjoyed sharing kumiss – fermented horse or camel milk – with the Mongolian team-members, and hunted wild game such as hare or the mountain sheep called *argali*.

They were not alone in the dry valleys. Herds of ibex and gazelles could be seen wandering the hills, and around the camp skulked lizards, jerboa and greedy long-eared hedgehogs that rootled through their rubbish. Not all of the wildlife was so benign. Marmots, they were warned, should be avoided because they carried the plague. Flies were a problem at one of their camps, the swarms thickening the air like soup. Poisonous vipers and scorpions had to be removed from their tents on a regular basis. But by far the worst were the enormous solpugids, or camel-spiders. Halfway between a spider and a scorpion, these hand-sized horrors lurked on the walls and crawled into their beds like insidious nightmares. The team hated them above all else.

Fieldwork revolved around the fossils and the heat. Each day the team ate at 7 a.m. and began work at 7.30 a.m. Just like my team on Skye – and Mesozoic mammal searchers the world over – they spent their hours crawling the Earth with faces in the soil, like truffle-hunters. Unlike in Scotland, they did it for weeks at a time under the fierce blazing sun. At 1 p.m. they stopped for lunch in the shade of their tents, remaining there until 5 p.m. to escape the blistering scorch of the afternoon. Then the team set out again, and didn't return until dinner at 8 p.m.

The temperatures were extreme. It regularly reached 40° Celsius during the day (100° Fahrenheit), and sometimes well above. One day it was so hot that when a rogue raincloud passed overhead, the droplets evaporated before they touched the ground. Sandstorms tore down the camp, leaving it strewn with debris. At night they sometimes slept outside under the stars, waking to find their sleeping bags caked in hoarfrost.

Their first complete dinosaur discovery was a young *Tarbosaurus* skeleton (a kind of tyrannosaur), poking out from the lower beds of a small cliff. They found many more, including the biggest specimens known at the time. Sauropods had previously only been found in pieces in Mongolia, but they recovered the first complete skull. They named new species of dinosaur, including *Deinocheirus*, and new types of *Gallimimus* – relatives of those known from North America. And they added to the specimens of *Protoceratops* and *Velociraptor* and ankylosaurs found by previous expeditions.*

Of course it was the much smaller mammals that really interested Kielan-Jaworowska. The team retraced the steps of the expeditions by the Americans in the 1920s, and others made by the Soviet Union in the 1940s. At Bayn Dzak – the 'Flaming Cliffs', so-called for their red sandstone that glowed scarlet in the sunset – they searched the slopes and recovered new mammal fossils. These were eroding in lumps from high on the cliff face, and tumbled down to their feet like dropped coins. Following the seams of rock for miles, the team uncovered new fossil-rich sites of their own.

The Polish-Mongolian expeditions continued until 1971, and by their end they had discovered the best specimens yet known from the Gobi. They extensively mapped the areas they worked, keeping meticulous records. Kielan-Jaworowska made sure that their work was carried out in concert with Mongolian researchers, and at the end of the season they donated most of their equipment to the newly founded Paleontological Laboratory in Ulaanbaatar. They trained local people in the field, including two women named Zorikht and Gunzhid who were among their most determined excavators. Later, Kielan-Jaworowska mentored and supported Mongolian researchers, always generous with her time and expertise. Many of the earlier

---

* These were not the first people to discover fossils in the Gobi and other parts of Central Asia, of course. The discovery of such bone fields by inhabitants of these regions over the last few thousand years may have been interpreted as the remains of animals we now know as mythical beasts, such as the griffin.

Russian studies from the 1940s were only published in Russian, limiting their reception internationally, but crucially, the Polish-Mongolian team published their discoveries in English in international scientific journals, making their sensational discoveries accessible to the established scientific world.

By the end of 1965 they had already found 30 fossils of new mammals. Mongolia once again took centre stage as palaeontology's El Dorado, and brought mammals from the time of dinosaurs to the forefront of evolutionary science.

⚬ ◆ ⚰

By the Middle Cretaceous, Mongolia had raised its head above the shallow northern fringes of the Tethys Sea. Although it had patches of parched scrubland, these were interspersed with fertile forests, brimming with lakes and criss-crossed by rivers. Stands of verdant forest fed and sheltered herbivorous animals, including dinosaurs and mammals. These were preyed on by predators of all sizes.

The fossils found by scientific teams in the Gobi have been written about many times. They include some of the enduring names recited by schoolchildren around the world. As well as skeletons, key among their discoveries were the first recognised dinosaur eggs[*] from the Flaming Cliffs.

The eggs were uncovered by the expeditions of the American Museum of Natural History in the 1920s,[†] which had piqued Kielan-Jaworowska's interest as a child. When Andrews and fellow expedition member William King Gregory returned from Asia to the United States, all anyone wanted to know about were the eggs. They were hounded by the press for images, information and quotes about them, fuelling renewed interest in dinosaur palaeontology.

Despite this later emphasis on reptiles, the key reason for the expeditions had actually been pursuit of our mammal origins. In a

---

[*] They were not the first dinosaur eggs ever found however. The priest Jean-Jacques Pouech saw 'eggshell fragments of very great dimensions' in the Pyrenees in 1859, he just didn't know what they were.[20]

[†] But they had been found by indigenous people in the area long before. Andrews described necklaces made of dinosaur egg shell in Neolithic sites near the Flaming Cliffs.

strange echo of the words of Buckland 100 years previously, Gregory
wrote in an article in 1927,

> ... the discovery of six imperfectly preserved fossil skulls of small
> mammals in the very same beds that yielded the dinosaur eggs was an
> event of exceptional importance only to the very few ... but on the
> paleontological side of the expedition the Cretaceous mammal skulls
> are perhaps the most valuable fossils so far discovered.[21]

These 'wee timorous beasties'* as he called them were the first of the
modern mammal lineage, the therian mammals. Just like the jaws
from Stonesfield, the fossils were eclipsed by their showy dinosaur
neighbours, but Gregory recognised their significance. One of the six
skulls was photographed held in the desert-dry hands of an expedition
member. In the background, out of focus, a tobacco pipe is gripped
between his lips. In the centre of the image is a skull about the size of
a cherry, gripped between thumb and forefinger.

The skulls were shown to the American inheritor of Marsh and
Owen's fossil mammal legacy, George Gaylord Simpson. In 1928 he
produced the most comprehensive catalogue of known Mesozoic
mammal fossils since Owen's day. Simpson is one of the most influential
palaeontologists in history, particularly for his seminal work *Tempo and
Mode in Evolution*, published in 1944. He pioneered the study of the
speed that evolutionary changes could occur and how this was related
to the way in which they happened. His ideas contributed to a wholescale
reworking of palaeontology occurring in the first half of the twentieth
century: the 'modern synthesis'. This new understanding of evolution
combined Darwin's original theories with new knowledge of genetics,
population dynamics and patterns of broad-scale evolutionary change
on geological timescales (known as macroevolution). It used mathematics
to analyse them, and prioritised data over individual observation.

---

* A reference to the best of poems by Scotland's national bard, Robert Burns, 'To
a Mouse': 'Wee, sleekit, cowrin, tim'rous beastie, / O, what a panic's in thy
breastie!'. Incidentally, this poem is also the source of the title of John Steinbeck's
novel, *Of Mice and Men*: 'The best-laid schemes o' *Mice* an' *Men* / Gang aft agley'
(= often go awry).

Simpson and Gregory worked on the Gobi mammal skulls together, coming to the conclusion that 'even in the Lower Cretaceous the mammals as a class were already far from the beginning of their career'.

The Asiatic expedition had found the forerunners of all modern mammal radiations. More searching was needed to fully understand what this meant for the origin of our modern world.

By the 1920s metatherians – the group that includes marsupials and their closest kin – had been found in Cretaceous rocks in the United States. Their sister group, the eutherians – including placentals and all mammals more closely related to them than to metatherians – were not yet confirmed as living in the time of dinosaurs, but they had been found in the Palaeogene, just after the Cretaceous mass extinction. This suggested that they had first appeared in the Cretaceous, but without the fossil evidence scientists could only infer our branch's earliest origins. What they needed was proof.

The first Gobi expeditions had found evidence that eutherian mammals existed in the Cretaceous. One fossil in particular, named *Zalambdalestes*, was not only thought to be eutherian, but also one of the most complete Mesozoic mammal fossils known at the time.

The differences between eutherians and metatherians in the Mesozoic lies foremost in their teeth: for example eutherians have three molars, while metatherians have four or more. The cryptic diagrams of tooth cusps that accompany publications on these differences look more like alien glyphs than anything biological, but they outline clear delineations in the arrangement of cusps in these early lineages. The bones in their ankles are also slightly different, as are details of their skulls and skeletons. From the outside however, both groups were probably quite similar to look at, resembling many small mammals we see around the world today.

It was thought from fossils now known from China, Mongolia and Uzbekistan, that metatherians could have originated in Asia. The oldest fossil purported to be metatherian, *Sinodelphys*, was a small possum-like tree-dweller found in the Jehol Biota. This site of exceptional preservation is similar to the Yanliao Biota, but dating to the Cretaceous rather than the older Jurassic. It contains an equally mind-blowing diversity of fossils, preserved in photographic detail. The presence of *Sinodelphys* suggested that before North America could finally pull away from Eurasia's embrace, some of these earliest metatherians had reached the New World. The first marsupial mammals were then thought to have further evolved there, before later journeying to South America and Australia.

More recent research casts doubt on this theory of metatherian origins. In 2018 a new specimen from the Jehol Biota allowed researchers to take a second look at *Sinodelphys*. They now think it is in fact eutherian, which means that the oldest undisputed metatherian fossils are from much later rocks in North America. Exactly where and how the group originated remains a mystery for the time being.

Alongside *Sinodelphys* in the rocks of the Jehol Biota was fellow climber and purported eutherian *Eomaia*, the 'dawn mother'. It is a furball: a circle of brown with toothpick bones laid around it. Your mother truly was a hamster.* This splodge of roadkill puts modern mammal lineages on the scene in Asia around 125 million years ago, well before the 66-million-year date that marks the beginning of the modern mammal story most of us are taught.

Suffice to say, the first members of the major modern mammal groups have good vintage. Because they are still alive today, we hunt through mammal evolution hungry to find them. We play *Where's Wally*† with each tableau, looking past a thousand other characters to spot them in the crowd (*Look, there he is! Turn the page!*).

But the ancestors of modern mammals were still only the understudies at this point in the story. There were more flamboyant characters on the Cretaceous scene with compelling tales to tell us about evolution, competition and success.

By the time they had finished their Mongolian fieldwork in 1971, 'the collection of Cretaceous mammals gathered during the expeditions', wrote Kielan-Jaworowska in an article describing the research, 'constituted the largest collection of mammal skulls from the Mesozoic

---

* A reference to the wondrous insults of a French castle inhabitant in *Monty Python and the Holy Grail*. As I've pointed out of course, these are not in fact rodents of any kind, because rodents hadn't evolved yet. I can't comment on whether they smelt of elderberries.

† Or *Waldo*, if you are North American.

Era anywhere in the world.'[22] The majority of their finds were not from therian stock, but belonged to other mammal groups that were far more successful at that time. Chief among them were multituberculates.

Kielan-Jaworowska set about studying the multituberculate skulls with the same eye for detail she'd used in her invertebrate research years before. She reconstructed the path of nerves and blood vessels three-dimensionally, using serial grinding[23] – the precursor of CT-scanning invented by Sollas.* She used her findings to show how these most successful of mammals were related to the rest of the family tree – including the therians – resolving long-disputed family relationships.

There were not only skulls to examine, but limbs and bodies. She noticed that bones in the pelvis once thought to be unique to metatherians and monotremes were also found in multituberculates, giving the first clues to their shape in the common ancestor of all mammal groups. The leg bones of some multituberculates suggested they hopped on all fours like a bounding rabbit, a surprisingly specialised mode of locomotion for an ancient mammal. Kielan-Jaworowska's research transformed multituberculates from a niche interest into one of the biggest subjects in Mesozoic mammal palaeontology.

The multituberculates were the most common mammals in the Cretaceous, comprising as much as half of all species. Many of the fossils from Mongolia have names suffixed with *baatar*, Mongolian for 'hero'. There is *Kamptobaatar*, the 'bent hero', *Tombaatar*, the 'big hero', and it was only natural that before long there was a *Zofiabaatar*, in honour of the woman who truly put multituberculates on the palaeontological map. This mammal group dwelled in the forests, grasslands and arid places of the northern continents. They climbed trees and dug. Some hopped through the dry scrublands like jerboa. There are species the size of beavers, and one group had a massive premolar tooth like a circular saw, which sliced up food ready for grinding – it may even have allowed them to splash out on the occasional meat dish.

In his 1927 article on the mammals of the Gobi, Gregory pokes fun at the dinosaurs, who for all their size and success barely noticed these creatures slowly taking over the world: 'thus the mighty were

---

* See Chapter 8.

put down and the meek inherited the Earth'. What their discoveries, and those of the Polish-Mongolian expeditions and recent Chinese specimens, have taught us is that mammals in the Mesozoic were already well established and thriving. They built on a long history of boundary-breaking innovation. From their miniaturised night-time bottleneck in the Triassic, mammals had gone on to quickly diversify in size, living in every habitat and exploiting niches once thought only the preserve of the post-dinosaur world. They were far from the 'meek' creatures that their piecemeal early fossils had once suggested.

Perhaps the greatest shock in the Mesozoic mammal story came at the turn of the twenty-first century. A discovery from Cretaceous rocks in China would prove that one mammal group were not only bigger, but also eating the most unexpected dish of the day.

They were the first mammalian[*] predators, and their dish of choice was dinosaur.

Being a dedicated flesh-eater is no straightforward route for survival. In the therapsids we saw the first large specialised carnivores in the synapsid lineage. They displayed obvious predatory traits like long sharp canine teeth for biting, bodies built for chasing and pouncing, and powerful necks and forelimbs for holding on to their prey.

But most carnivores are not exclusive meat-eaters. Although we use the terms definitively, animals rarely feed on one food type alone. Those that do may struggle to survive when extinction or habitat loss hits their target chow – giant pandas (*Ailuropoda melanoleuca*) are the perfect example of this.[†] As little as a third of a carnivore's diet might comprise meat. Most bears for instance rarely eat more than this, having flesh as a top-up to their core diet of berries, nuts, roots and shoots. Most members of the dog family, such as foxes, split the

---

[*] They were true mammals, members of Mammalia.

[†] The giant panda's taste for bamboo, a fantastic way to survive in a region riddled with bamboo plants, now restricts them to select regions of China. Their habitat is being lost to deforestation, turning their specialism into a disadvantage in the face of rapid human development, and making them the monochrome poster-child for threatened species.

bill: with around half of their meals meat-based, with vegetable matter and even fungi forming the other 50 per cent.

Then there are hypercarnivores. These animals have diets comprising at least three-quarters meat. When we think of them our imaginations fill with roaring tigers and *T. rexs*, but we ought to calm down a little. Hypercarnivory isn't necessarily so sexy: sea-stars and axolotls* are also hypercarnivores. Although meat-eating animals can feed exclusively on living tissue when conditions allow, for the truly dedicated hyper- (or 'obligate') carnivores, it is their main diet. Their physiology and anatomy is specialised for processing high volumes of flesh.

Cats are obligate carnivores. They have simple, relatively short digestive tracts that lack the swarming symbiotic soup of fermenting bacteria found in herbivores. Feline guts are bastions of solitude – after all, the last thing you want flesh to do in your stomach is ferment. Cats need a lot of protein to get the nutrients required to survive, which is okay because there are always more herbivores around to eat than there are carnivores to eat them: that's food chain 101.

In the Cretaceous, multiple mammal lineages were hypercarnivorous insectivores. We know this based on their teeth and body size. Carnivores that eat other tetrapods tend to be larger in order to deal with their prey – although there are exceptions. The weasel family, long-bodied meat-eaters that include polecats, stoats and ferrets, show us just how ferocious a cute little slinky can be. The smallest member of the family, the least weasel (*Mustela nivalis*), is native to Eurasia, North America and Northern Africa, but has been introduced to islands around the world, including New Zealand. Despite being small enough to sleep in your palm, the least weasel can kill animals 10 times its size. It has proven devastating to indigenous wildlife in ecosystems where it has been released.

Sharp teeth and a big body are not the only evidence in the trial against one family of murderous Mesozoic mammals. A fossil from Liaoning Province in China, announced by Chinese researchers in 2005, became famous not only as the biggest Mesozoic mammal known, but for the incredible remains of its last meal preserved in its stomach.

---

* A type of salamander from Mexico with feathery gills. Very cute.

The fossil is not a pretty one. On my trip to IVPP as part of the Symposium on Mesozoic Mammals, I got to examine it first-hand. The mass of yellowed bones have been removed from the eggy-beige rock and set into plaster for easier study. The skull is smashed. Its spine runs the straight length of the specimen, with a fan of ribs visible and some messed-up limbs cluttering the extremities. Exhibit A in the case against the beast is a cluster of gunk in its midriff. Looking closely, the research team who described it saw that there were bones there, but they were not mammalian. They were the torn-up limbs of baby dinosaurs.

This new species of mammalian dino-diner was *Repenomamus robustus*. It was not caught red-handed, but full-stomached; its last meal was takeaway *Psittacosaurus*. This species of *Repenomamus* was the size of a North American oppossum, but its sister, *R. giganticus*, was as big as a honey badger, around a metre (3ft) from snout to tail-tip, and weighed up to 14 kilograms (31lb). They belonged to the carnivorans of the Mesozoic, a group called gobiconodontids.

Gobiconodontids and their closest kin were found all over the world, including cousins in Argentina, Tanzania and India. They lived from the Early Jurassic to the Late Cretaceous, when their numbers eventually petered out. Many were dedicated meat-eaters of insects or other flesh, but even the insect-eaters are no walk in the park. Take *Spinolestes*, the porcupine-like digger from Spain. It is preserved in the Lagerstätte* of Las Hoyas, an exceptional Cretaceous fossil locality. Incredibly, as well as its soft ears (see the previous chapter), its internal organs have also been preserved in the stone, including the lobes of the liver and a thin film of the muscular diaphragm. But it is the animal's coating of quills among a pelt of guard hairs that makes it remarkable. This is the first example of such armoury in a mammal. Those porcupine-like spines kept it safe as it feasted on ants and termites. Even the insect-eating gobiconodontids were a force to be reckoned with.

It's hard to know how much gobiconodontids actively hunted, versus opportunistic scavenging. Even the most adept hunters will eat from a carcass if the occasion presents. There are some dinosaur bones with small scratches in them, thought to be the gnawing of

---

* A lagerstätte is a rock deposit in which fossils are preserved exceptionally well, often with soft tissues like fur and skin.

opportunistic Mesozoic mammals. What we know for sure is that gobiconodontids had the dental equipment for dealing with prey, and although not all of them were large, they included the largest mammal members of the ecosystems they inhabited.

Stomach contents are preserved in other fossil specimens from the Cretaceous Jehol Biota in China. As you might expect, mammals feature on the menu for some smaller dinosaurs, such as the compsognathid *Sinosauropteryx*. This dinosaur looked like a skinny chicken with teeth, claws and a tail. There is a specimen with the jaws of the multituberculate *Sinobaatar* and a second mammal called *Zhangheotherium* inside its stomach, showing that they preyed on mammals (and other small tetrapods) as you might expect. It is likely that larger mammals also preyed on their smaller mammal relatives, as well as eating lizards and amphibians – and dinosaurs, it would seem.

Repenomamus giganticus remains the most complete and largest species of gobiconodontid known. But isolated teeth found elsewhere suggest there could be even bigger bad-asses just waiting to be discovered. The meek may have inherited the Earth, but there was far more to fear in their Cretaceous world than just dinosaurs.

◊ ◊ ◊

Our understanding of mammals in the Cretaceous – and the rest of the Mesozoic – is not yet complete. The furry paws of gobiconodontids, multituberculates and therian mammals clambered across the landscapes of the northern continents, but the evolutionary history of animals in the southern hemisphere is less well known. It is a problem for many time periods in Earth's history, leaving our knowledge of evolution lop-sided. There are multiple root causes for this.

Palaeontology has its historical origins in Europe and North America, so people there have been specifically looking for them for longer. Most of the rest of the world's treasures were funnelled into the former empires of Europe, who financed themselves with the plundered wealth of other nations. This wealth gave the predominantly upper-class men who first studied fossils plenty of time to dedicate to their pet projects and scientific pursuits. Fossils in their colonies were taken and hoarded in museums along with the rest of the loot. More often than not the 'first' fossil finds made outside Europe came long after their initial discovery by indigenous peoples. The knowledge of local communities was usually either ignored or exploited in the name of science. The instability created in countries finding their feet after a history of exploitation can make palaeontology a low priority, and sometimes dangerous. Addressing this legacy of empire is one of the most important challenges facing modern museums and researchers.

The pattern of discovery is also geographical and geological. Fossils can only be found where there is suitable outcrop of the right age. The proportion of land to ocean is lower south of the equator, and the majority of the rocks that make up our Earth's crust are covered with soil and vegetation anyway, preventing access. The lungs of the world, the Congo and the Amazon forests, are so densely vegetated that reaching fossils there is often virtually impossible. Antarctica only rolls up her glacial trouser leg enough for us to peek at her ankle rocks.

Many fossils are found when building infrastructure, which may not be present in harsher landscapes or countries facing political and socio-economic difficulty. With fewer roads, outcrops can require complicated logistics to reach and are a headache to retrieve. And of course it is also the case that rocks have been buried, tilted, tipped and eroded, exposed, inundated and washed away. Humans have built their cities on top of the puzzle pieces. Seams of evolutionary gold might run beneath your town square. Some of them form the very building blocks of architecture.

The palaeontological imbalance is being corrected. China is the ultimate example of a country that fully funds and explores its world-changing fossil heritage. The results in the last 20 years alone are spectacular, altering our understanding not just of mammal evolution, but also the origin of birds, amphibians – and of complex life itself.

Mongolian research today is led by local experts like Bolortsetseg Minjin. She was once only allowed to cook for Western field teams who visited her country, but is now the founder of the Institute for the Study of Mongolian Dinosaurs. She continues to fight for the repatriation of her country's heritage, having organised the return of over 50 dinosaur fossils and collections taken illegally by smugglers, who sell them abroad at auction for profit.[*]

---

[*] One of the most famous was the skeleton of *Tyrannosaurus bataar*, bought by the actor Nicholas Cage in 2012 for $1 million.

In South America the number of fossil localities is growing, and Argentina is particularly rich in Mesozoic outcrops. Some of the most exciting new cynodont fossils are being found and described by researchers there and in Brazil. In Africa discoveries have previously centred on South Africa, but new localities in Tanzania, Madagascar and Morocco are among those altering our understanding of animal distribution in the deep past.

As I write this chapter, a Mesozoic mammal from Madagascar has hit the headlines. It is named *Adalatherium*, the 'crazy beast', and belongs to one of the least well-known Mesozoic mammal groups. They are called gondwanatherians, and until 2014 no one knew much about them except for the scant presence of some enigmatic teeth. Now two near-complete skeletons from Cretaceous rocks on the island are beginning to tell us more.

The gondwanatherians have been found in Argentina, Madagascar, Tanzania and India. Like the multituberculates they made it through the asteroid impact, with fossils from South America and Antarctica attesting to their continued presence in the Palaeogene. *Adalatherium* lived just prior to the extinction event, by which time Madagascar had been isolated from the rest of Africa and India for around 20 million years. As a result it is a unique creature. It resembles a giant groundhog, complete with the pincer-skull that served the groups like the tritylodontids so well for millions of years. Comparing it to other mammals in the Cretaceous, it appears that gondwanatherians might be a close sister-group to the prolific multituberculates, or even an offshoot along the same branch. Others suggest they're related to the haramiyidans, the group we met in the previous chapter who learned to glide in the forests of the Cretaceous.

There is still a great deal to learn about the evolution of mammals in the southern continents. Without a doubt these are the countries that hold the future of palaeontology in their hands.

Zofia Kielan-Jaworowska's influence transformed Mesozoic mammal research. As well as over 220 articles and books, she published more than 70 popular books and articles for young and old, bringing early mammals to a wider audience than ever before. She helped organise and run conferences, museum exhibitions and field work. She fostered talented new researchers from every continent, including Mongolia and China.

Yet she remains one of the many under-acknowledged women of scientific history. In a tribute after her death, John R. Lavas ranked Kielan-Jaworowska alongside Marie Curie as one of the most important and influential female scientists of all time.[24]

In photographs taken in 1974, Kielan-Jaworowska stands on the plaza outside a meeting in Moscow. Beside her is fellow female scientist Nina Semenovna Shevyreva, a specialist in rodent palaeontology. Flanking them are middle-aged men in dark suits. Kielan-Jaworowska is laughing, her dark hair caught in the wind and white dress glowing. In another image at a party she stands next to the ageing G. G. Simpson, deep in conversation. In the Gobi desert she is one of only three scuff-kneed and sun-brown women in a line-up of shirtless male torsos. In her most iconic photograph, she lies belly-down in the sand of the desert, engrossed by a fossil fresh-plucked from the russet earth.

By the time her biblical compendium of Mesozoic mammals, *Mammals from the Age of Dinosaurs*, was published in 2004, Kielan-Jaworowska had returned to Poland after a long career abroad, much of it at the University of Oslo. She wrote the book with two young researchers she mentored who are now among the leaders of the next generation, Richard Cifelli and my own tutor, Zhe-Xi Luo. She was almost 80 years old, but like all of the greatest scientists, still hard at work. Her compendium remains the only publication of its kind, outlining every Mesozoic mammal known at that time. It bows the bookshelves of palaeontologists around the world under its heft.

In 2011 her husband and love of her life, Zbigniew, passed away. She threw herself into her work. Her home just outside Warsaw became an epicentre of palaeontology. Kielan-Jaworowska welcomed scientists and friends from around the globe to her home office.

She died in 2016,* still sharp and astute to the last. Hers was a life marked by revolutions: those she endured in life, and those she incited

---

* I wish I could have met her, but as my career was beginning, hers ended. At the Mammal Symposium in China just a couple of years later, I sat at a table with several people who knew her well, and we drank Baijiu as they told stories about her remarkable life and work. Those people now pass on what they learned from Kielan-Jaworowska to the next generation of students, and so we continue to benefit from her sharp mind and great generosity.

in science. She should be remembered as one of the greatest scientists of all time.

◇◇◇

The Cretaceous mammals studied by researchers like Kielan-Jaworowska thrived thanks in no small part to the Cretaceous Terrestrial Revolution. Mammals still retain an intimate ecological link with flowering and fruiting flora. Not only did this direct link shape mammal evolution, but the changing plant life influenced insect evolution too, providing new food sources for insectivores as well. There are around 350,000 flowering plant species known in the present day, making up 90 per cent of all plants and vastly outperforming the rest of the plant world in terms of diversity.[*] So what was it that drove their sudden diversification in the Cretaceous?

In Darwin's day scientists pointed the finger at pollination. They suggested that pollinating insects and flowers adapted together. They theorised that the flowers became increasingly attractive for pollination, and in turn insects dispersed the pollen more effectively and relied on the flowers more heavily. This idea certainly explains the increasingly elaborate and co-dependant relationships between flowers and their pollinators (which can include mammals and birds as well as insects). But it fails to explain why angiosperms took so long to get off the starting block.

A study in 2009[25] sought answers not in the flowers themselves, but in their leaves.

Leaves breathe. They inhale carbon dioxide, releasing oxygen from water in exchange. These materials are transported through a system of veins. The more veins they have in their leaves, the more rapidly they can make this exchange and the more productive they are. A team of researchers from the United States and Tasmania decided to measure the number of veins and pores in fossil leaves to look for patterns and work out how lushly they grew in deep time.

---

[*] But not necessarily in terms of abundance. There are only around 1,000 gymnosperms, but they tend to dominate vast sections of the landscape of the Earth, such as the boreal forests, which are mostly coniferous.

Amazingly they found that the number of veins in angiosperm leaves increased by an order of magnitude in the Cretaceous. In the last 400 million years of plant evolution, despite fluctuating climate conditions and atmospheric composition, no other plant group has even come close to the irrigation seen in flowering plants. Although this doesn't answer things once and for all, it does suggest that angiosperms in the Cretaceous may have found themselves suddenly able to outperform gymnosperms in terms of growth. Their increased transpiration may have created their very own moist microhabitat, paving the way for a lush understorey unseen in coniferous forests. If pollination and the development of tasty fruit added to their capacity to disperse seeds, the combination might have given them an edge.

The Cretaceous Terrestrial Revolution wasn't good news for everyone. This turnover in flora seems to have given rise to a multitude of new niches, resulting in an increase in animal diversity. But the tritylodontids, stalwarts of the Jurassic, were among the losers in this shifting world.

It has traditionally been assumed that tritylodontids became extinct because they were out-competed by the multituberculates. The idea that one group can usurp another is also called the 'competitive exclusion hypothesis'. This suggests that no two groups can coexist to exploit the same niche in nature – the weaker one (that is, the less well adapted) will always eventually be replaced by the stronger. It has also been called Gause's rule, after the Russian biologist who tested the idea in the early nineteenth century. This simplistic explanation ties into the classic military language of evolutionary empires rising and falling, a language of combat and war. The truth is more complex, and less to do with strength than luck.

Looking at populations mathematically and in the laboratory, one species can easily replace another if they are better at exploiting resources (food, water, light and space). But when ecologists turn to the natural world, they find populations appear to flagrantly flout the rules. Species that feed on the same resources can overlap because ecosystems are too complex to explain through sets of rules and numbers.

Interactions between animals and their world can change from place to place and year to year, affected by climate and the distribution of competitors and habitats. Specialists may emerge to take over a niche, then lose out and die off as conditions change again. Over longer time frames populations fluctuate, hit by anything from disease to sun-spot activity. The system is so multidimensional that the study of nature – especially in the geological past – remains ever vibrant and baffling. Complete understanding is always just beyond our grasp … but close enough that we continue to reach.

It's easy to assume multituberculates were simply superior to tritylodontids. Perhaps the mammalian ability to feed underdeveloped young on milk increased the survival of litters. Or maybe their diphyodont teeth chewed food more efficiently, providing more nutrients faster. Multituberculates were 'true' mammals after all, they must have been 'better' than their predecessors … If we're not careful we can find ourselves leaning back into the old Victorian rut of evolutionary progression.

At present, there is very little evidence to support the idea that multituberculates directly competed with and excluded tritylodontids. But the timing of the shift – coinciding with the revolution in flowering plants – hints that the changing composition of the greenery might have given multituberculates an advantage. Tritylodontids were as good at the game of life as any mammal, it's just that nature changed the rules again.

Soon a change came that threw the rule book out of the window altogether. A much bigger shift in circumstances would confront not only our mammal family, but the whole of life on Earth.

CHAPTER ELEVEN

The Journey Home

Bemoaning thus, by dumous path, they crushed the cycad's growth,
And many a crash, and thunder, marked the progress of them both.
And when they reached the estuary, the excandescent sun
Was setting o'er the hefted sea; their saurian day was done.

Ethel C. Pedley, *Dot and the Kangaroo*

I could not sufficiently wonder at the intrepidity of these diminutive mortals …

Jonathan Swift, *Gulliver's Travels*

In a nest furnished with moss, a small circle of fur twitches. Nose tucked against warm belly. Whiskers flickering. The sleeper is thrilled by memories of herbal scents and squirts of crushed bug. Senseless dreams of scampering curl its minute toes. The burrow is Goldilocks warm, just big enough to turn around and tuck in your tail. It is mushroom dark, filled with musk. The Earth clasps this sleeping body in her hand.

The little sleeper barely stirs as a rumble reaches it from another continent. No different from a herd of dinosaurs passing overhead. It drifts off again. Hunger brings it round several hours later, and it scurries off like a Tube train through the tunnel network. It picks up beetles and chews a few stray rootlets. Back to central station – groom, stretch, sleep again. Overhead a layer of soot smudges across the sky, then falls like snow.

Days later when this small mammal finally surfaces, it runs the usual rounds. Scent messages are left for others to sniff, paths wind under logs and into crevices, avoiding the layer of ash. A large dead animal in a nearby clearing has formed a temporary metropolis of insects and small-bodied scavengers. The mammal scoots in for a takeaway.

The air is alive with flies. Months have passed and the small mammal grooms its glossy coat ready for the mating season. There are plenty of partners to select from; its numbers have exploded. They are well fed on a sudden glut of carrion, with side dishes of fungi. They chase and squabble in the open – fearless this year with so few predators to pick them off. Is it day? Is it night? Does it matter when your eyes are filled with sensitive cells for semi-darkness?

Four pink wigglers writhe in the moss. The mammal cups them with her fuzzy body and licks their hot skin. It's unnaturally cold outside, but the burrow stays the same, a quiet haven for her litter. As long as she can find enough snatched mouthfuls to feed herself, she can convert it to nourish her babies. They suckle milk filled with nutrients foraged from the death spreading above.

The landscape languishes, littered with bones as winter lasts for years. The sky begins to clear and mounds of bodies melt under the returning sun. A sweep of ferns shifts the landscape back to green. The mother mammal is gone but generations of her offspring continue their daily work. New tunnels expand. A hundred generations of pink wigglers grow up to inherit the Earth.

◌ ◇ ◌

From a mammal's point of view, the epic end-Cretaceous catastrophe might be nothing more than a harsh season to weather. It's not that mammals were unaffected – like most animals they suffered catastrophic casualties. Many lineages became extinct. But we all know our own creation mythology: our ancestors rose like a phoenix from dinosaurian ashes.

The mass extinction that killed off the non-avian dinosaurs is called the K–Pg event: Cretaceous to Palaeogene. It happened around 66 million years ago. The doctors Alvarez (father and son) famously headed the geological team who discovered a layer of rock marking this geological transition. The universe drew an iridium line under things. From here on, life would be different.

Atomic number 77, iridium is a rare element on our planet. It is the stuff of asteroids. The abundance of it sandwiched in the K–Pg sediments suggested that one had hit the Earth. In 1980 the Alvarezes proposed such an event had occurred and left trace remains scattered over the planet's surface. Molten rock spherules and shocked quartz in the same layers supported their conclusions. The hunt began to find the crater.

A decade of searching led to the north coast of Mexico. Although an impact big enough to kill up to three-quarters of all species should have been unmissable, it was tricky to spot because it lay at the very edge of the Gulf of Mexico. Half of the evidence was underwater. But what scientists traced matched the size estimates for an impact crater.

At around 150 kilometres (almost 100 miles) from rim to rim, it would take a couple of hours to drive across it on a good road, if such a thing existed across its diameter. The crater was named Chicxulub after the town that now sits almost in the epicentre. This rocky pock-mark is 20 kilometres deep (12 miles). Maps of the subsurface structure reveal rings of rock, like a photograph taken the moment after a water droplet pierces a pond's surface.

There are still those who argue that the impact which gouged out the Chicxulub crater is not the cause of the K–Pg mass extinction, but the evidence is quite overwhelming – it is safe to say any jury would convict. Rock-dating methods confirm the timing of the impact. There may have been other contributing events, however. At the end of the Cretaceous sea levels were falling, which may have driven an increase in natural extinction rates. Trap lavas – similar to those in Siberia that destroyed almost all life at the end of the Permian – flooded India at around the same time. They undoubtedly contributed to local extinction, and their aerosols to global climate change. When the K–Pg impact came it may have found life on Earth already facing hard times.

The asteroid that forged the Chicxulub crater was the size of a small island. Imagine part of the Hawaiian archipelago dropping from the sky, or the Isle of Skye, Cuillins and all. The impact was several billion times stronger than the Nagasaki and Hiroshima bombings. An earthquake larger than any recorded in human history would have made the Earth reverberate like a bell. The thermal shockwave would have flash-fried all life for hundreds of miles. The blast of air probably flattened forests as much as 1,000 kilometres away.

The touchdown location was about the worst possible place to gut-punch the Earth. By landing in the shallow sea the asteroid displaced enough water to create a mega-tsunami at least 100 metres (330ft) in height – some estimates are closer to 1,500 metres (5,000ft). This wall of water mounted the coasts of North America and barrelled inland like a liquid steam-roller. Across the Atlantic Ocean a smaller tsunami careened into the coastline of Europe and Africa; despite the distance from the epicentre, this wave would still have been the height of a three-storey house.

The rock on the edge of the Gulf of Mexico contains large amounts of gypsum, which is full of sulphur. The impact vaporised this, sending a splash of liquefied rock into the atmosphere. Ejecta hot with radiation would have rained down for hours. Life over as much as half of the planet would have been instantly affected, with billions dying within the first day. The wildfires that scorched the rim of the Gulf of Mexico would have made today's forest fires look like pleasant summer barbeques.

The dust that took residence in the atmosphere maliciously swirled its way around the planet, carried on air currents and jet streams, until it enclosed all life in its smothering grip. Each day the sun still rose, but as little as half of its light could penetrate the dense weave of aerosols. For a year or more, those left alive struggled through a daily cycle of darkness and simmer dim.[*] The sulphur in these sooty pillows combined with water droplets to make sulphuric acid. This spiteful liquid threw itself into the face of the Earth. The lush green top layer was burned away. Soon even the herbivores most distant from the impact site languished.

A nuclear winter blew in and lasted for years. The sudden alteration of surface temperatures whipped up storms, battering the survivors. Carnivores might have managed for a while, picking off wandering plant-eaters and scavenging on carcasses. But these supplies would only have lasted so long. The complex food webs of the Cretaceous were disassembled, the parts removed piece by piece until only a ruinous outline of an ecosystem remained.

---

[*] A Shetlandic term for the light at midnight in midsummer, when the sun barely sets. Shetland is the northernmost place in Scotland. Further north, the sun doesn't set at all in midsummer, and is known as the midnight sun.

An extra few hours, and things might have gone quite differently. If the Earth had turned its cheek the asteroid could have collided with the Pacific Ocean or the young Atlantic. Although the volume of water displaced by the impact would have been greater had it hit the ocean, the deep sea would have absorbed the shock. There would have been less rock ejected into the atmosphere and less sulphur, reducing the long-term damage of the impact. Non-avian dinosaurs might have made it.

Few animals bigger than a Labrador dog survived the K–Pg extinction event. The small body-size of animal populations after mass extinctions is called the Lilliput effect.* Larger animals may be especially badly affected by cataclysms and unable to find the resources to sustain them. They may also reproduce too slowly to recover. Those that do survive could find themselves faring better with smaller body masses that require less energy to sustain, and so natural selection quickly shrinks them.

The K–Pg was a guillotine blade to the dinosaurs. There is little sign that any non-avian dinosaurs survived long, if at all. Any that did were what palaeontologists call 'dead clade walking', hanging on for a brief time but never recovering. The earliest bird groups, which appear to have been doing well prior to the impact, were lost, as were many of the Cretaceous branches of modern groups.[1] It may surprise you to learn which birds made it through: the ancestors of ducks, fowl and ground birds like the emu accompanied a few small-bodied flying birds, populating the new era.

Pterosaurs appear to have already been sieved out of the Late Cretaceous world. Only a few remained, notably including giraffe-sized giants, the azhdarchids. It has been suggested that pterosaurs may have been replaced in the majority of their niches by the steadily expanding diversity of early birds, although this theory suffers from the same caveats as the tritylodontid-multituberculate replacement debate from our previous chapter. When the asteroid came however, the Titan azhdarchids perished.

---

* Named for the tiny island nation in Swift's *Gulliver's Travels*, where the people are a twelfth of the size of regular humans.

Being smaller didn't save everyone – multiple lizard[2] and amphibian groups became extinct. And of course, this disaster hit mammals hard too. The prolific multituberculates did poorly in places like Asia, as did the metatherians, the group that includes marsupials. Plants felt the impact, and the insects that relied on them. With dust particles and aerosols cutting light and radiation from the sun, leaves struggled to photosynthesise and died en masse. A spike in fungal spores preserved in sediment from the K–Pg betrays the widespread rot,[3] as it did in the end-Permian extinction long before. Bees suffered as their food sources were diminished,[4] with many species lost.

In waterways and seas crocodiles were hit hard, and marine reptiles and other organisms disappeared or were severely affected. The ammonites – perhaps the best recognised and most commonly collected fossils on the planet – vanished from our seas for ever, remembered for the rest of time by their countless empty spirals.

We can't know exactly when the very last member of any of these groups was blinked out because of the Signor–Lipps effect. Despite the sound, it's not the stage name of an Italian soft porn actor, but a palaeontological rule when reading the fossil record.[*] It states that we'll never know when any species appeared or became extinct, because we'll never find the fossil of the very first or very last individual. Instead, we can only get the minimum and maximum range, an error bar of bones to constrain their existence.

Some estimates suggest that the mass extinction itself only lasted a few hundred to a few thousand years. Others say tens of thousands. With certainty we can say it was a geological camera flash.

Recovery from the extinction would have varied from place to place. Studies from Colorado hint that ecosystems were beginning to repair themselves within half a million years. Within four million years, leaf-eating insects in places like Patagonia had recovered.[5] Although some of the associations between plant and insect had ended, novel new arrangements were struck that underpinned the reborn ecosystem. Ants and termites took on a new importance and flourished, as did butterflies. Angiosperms recovered quickly. They

---

[*] Named for the researchers who proposed it, Philip Signor and Jere Lipps.

gave rise to new lineages of fruit which proliferated in the aftermath, taking advantage of the increasing number of hungry mammals and birds to disperse their seeds.

The deepest oceans remained largely untouched by the devastation. Acid rain had poisoned shallower seas and weakened their calcium-shelled inhabitants, but it wasn't enough to alter the dark depths of the water column. Between the plants and oceans there was a sturdy scaffold to erect new food chains.

For some animal groups it was less an extinction than a re-shuffling. As in other mass extinctions, specialists were hit hardest. It took flexibility to make it through, an evolutionary make-do-and-mend. Those that could cope with the stress of the immediate impacts of the event rode it out – perhaps even flourished for a time. Some species clung on in refugia, places less affected by the changes taking place in the wider world, like ecological Shangri-las. They emerged blinking in the sunlight, to reclaim the emptied lands.

As the dust settled on a post-dinosaur world, it was the eutherian mammals who emerged first and fastest. This group had been mere bit players in the second half of the Mesozoic. Their meteoric recovery in the Palaeogene would have repercussions for the whole of life on Earth.

It was time to begin the second 'age of mammals'.

◇ ◇ ◇

The Palaeogene was a bit of a mess. Like the aftermath of any storm, it took time to rebuild. This time period began explosively 66 million years ago, and lasted 43 million years until the Neogene began. Together the Palaeogene and Neogene used to be known as the Tertiary, following faithfully from the Secondary rocks of Victorian parlance. Although this name is still used occasionally, it has become defunct as our geological timescale has been refined.

After the Neogene, the baton was handed to our current geological division of time, the Quaternary, which began two and a half million years ago. The Palaeogene, Neogene and Quaternary are together called the Cenozoic. We are just a blip at the end, a fingernail on the arm span of vertebrate life.

It has been challenging to retrace the evolutionary scramble among mammals to get their act together in the Palaeogene. Many books will tell you that mammals began there and diversified from small and simple ancestors. We now know that this is not true. They had a complicated heritage, and mighty beginnings that stretch 300 million years further back in time.

However, there is *some* truth to the stereotype of modern mammal origins.

What is it that makes mammals and birds so special? The short answer is that they are not. Many lineages made it through the K–Pg extinction event as we've already seen. Birds are pretty prolific in the modern world, but one glance at the number of fish species and you'll re-adjust your parameters for diversity.* For some reason humans are just mad keen on these two groups, inviting them into the garden, making films about them, even adding them to the family like long-lost offspring.

But there are some interesting parallels to draw between mammals and their feathered compatriots. Some of these similarities may have contributed to their fast re-emergence from the extinction event.

Because of their significance to us, we've dedicated a lot of time to understanding how mammals proliferated after the K–Pg. One of the most pressing questions is why our furry ancestors didn't diversify sooner. What kept us in the supporting roles for so many millions of years? It has long been assumed that it is the presence of dinosaurs that held us in check. It seemed obvious from the fossil record, but until the end of the twentieth century there was very little solid data to support it.

Researchers began crunching the numbers. A study in 2013 really tested the data to the max. Using techniques that would have made G. G. Simpson proud, Graham J. Slater explored the speed and pattern of changes in mammal body mass through the Mesozoic and into the Cenozoic.[6] We know that being small is not a measure of success, but it is a useful measure for how morphologically different animals are from one another. A change in size represents a change in phenotype, the observable characteristics of an organism. By charting how phenotype changed in mammal families through time, Slater could

---

* If you remember, in Chapter 4 we discovered there around 5,500 mammal species, up to 18,000 birds and over 28,000 fish. There are over 5.5 million insects though, so it has been and always will be the age of insects.

compare his data to different patterns we would expect in different evolutionary scenarios.

There are several scenarios, or models, that mammal evolution might have followed in the last 200 million years. The first is Brownian motion, which is basically random. If this was the case we would expect the plotted data to fan out all over the place, with no discernible trend or pattern. This is what we would see if there was nothing stopping mammals from changing their body size.

The second possibility is that mammals were on their way up all along. This is the directional scenario, and the data would show a clear trend from small Triassic forms, steadily changing the range of their body size – at least in some lineages – through time in a rather neat pattern.

The third scenario is called the Ornstein–Uhlenbeck model.* If mammals evolved following this pattern, you would expect the plotted data to wiggle a bit, but not stray far from an 'optimum'. In other words, mammals would remain constrained within a certain boundary of size.

Slater found that mammals in the Mesozoic, despite a few dino-crunching exceptions, generally followed the Ornstein–Uhlenbeck model. After the K–Pg however, the scenario suddenly changes to Brownian motion. This step-change is best explained by a post-K–Pg release, followed by a radiation. In other words mammals were constrained, then experienced an adaptive radiation in their evolution which began 66 million years ago. This study provided quantitative data to support a long-established paradigm: that mammals were constrained during the time of dinosaurs, and the disappearance of all of them except birds at the start of the Palaeogene triggered a step-change, allowing them to diversify into the unoccupied niches.

It was an intriguing finding. Although this study may have confirmed the overall pattern of mammal body size evolution, it didn't really address the cause of the constraint in the Mesozoic. What else could it be but those dastardly dinosaurs, standing in the way?

---

* Named after two physicists, Leonard Ornstein and George Eugene Uhlenbeck, who worked out the maths that describe it.

It always seemed to me that the idea of mammal suppression was a little too simplistic. The first cracks in this cliché came with the realisation that mammals, far from being relegated to cowering in the dark, had adapted to it with such flair that it triggered wholescale changes in their brains and sensory systems. They were diversifying in a way not easily captured by analyses of body mass. More recently the discovery of docodontan diversity rewrote the narrative about what very early mammals got up to in their Jurassic ecosystems. With a steady increase in fossils from around the world – like *Repenomamus* and the gliding haramiyidans – we re-evaluated the limits of their body size and recognised their important place in the ecosystems of the Mesozoic. The whole story was being revised. Mammal groups had been radiating and experimenting multiple times throughout the time of dinosaurs.[7] The idea that dinosaurs had been keeping them in check has begun to look a little less self-explanatory.

Recently my colleagues and I took another look at mammal evolutionary patterns since the Triassic. Instead of focusing on the radiation of mammals after the K–Pg, we looked instead at the evolutionary constraint they faced in the Mesozoic. Aware of their diversity, we broke our analysis of mammals into three different groups: early diverging mammaliaforms (like morganucodontans and docodontans); the earliest 'true' mammals (such as multituberculates and gobiconodontids); and then the therian or modern mammals. We compared how much change took place in the phenotype of each group through the Mesozoic. The results were interesting, to say the least.

We found that it was the mammaliaforms and early mammalian branches that were most diverse for much of the Mesozoic, not the therians. These early groups continued to adapt into new niches long into the Cretaceous. Therian mammals, on the other hand, didn't get much done until after the K–Pg extinction event, hanging around backstage like understudies. This suggests that it may not have been the dinosaurs who kept our ancestors in check, but their mammal brothers and sisters.* The extinction of competing mammal groups at

---

* At the time of writing this study hadn't been published yet – fingers crossed by the time you read this it will have been.

the K–Pg probably played at least an equal role in the rise of modern mammals to the removal of those pesky reptilians.

We've already seen that body size is an important part of evolution. Synapsids were the first megaherbivores and giant predators in the Permian, marking the first 'age of mammals' over 250 million years ago. But despite them making it into the Triassic, it was the reptilian lineages that were fastest off the mark after the end-Permian extinction, and grew to giant sizes in the Mesozoic.

They may have been good at growing large, but it has to be said that most dinosaurs were rubbish at being small. We've been so focused on the gargantuan among them that until recently it wasn't really recognised that there is another side to the coin. A study in 2017[8] examined changes in body size among dinosaurs in the Mesozoic, and suggested that after reaching larger sizes, most lineages hovered around an optimum. This is the Ornstein–Uhlenbeck, where values never stray far from a certain range. For dinosaurs, this optimum remained relatively high.

However, abrupt shifts did occur from time to time through dinosaur evolution. This produced a few extra-large bodied lineages, and some smaller ones. The most important and radical of these happened in the theropods, particularly the ancestors of birds. Only the bird branch approached the tininess that mammals had achieved so long before them. Similar physiological changes fuelled their evolution: acquiring a body covering (feathers), increased and sustained body temperatures, and raised activity levels. Many of these changes occurred in the stem of their evolution, but they coalesced in birds, much as they had in mammals. Nature had set off in a different direction, but ended up arriving in the same place.

It is thought that this combination of traits might have helped birds and mammals to survive the nuclear winter of the K–Pg mass extinction event. Both lineages also care for their young more than most other tetrapod groups, giving the next generation a leg-up when times were tough. Habitat use likely played a role too: mammals that were semi-aquatic or that burrowed may have been able to escape the worst effects. It can be no coincidence that shoreline and semi-aquatic birds seem to have likewise done well. Studies of the few postcranial remains of mammals from the earliest Palaeogene suggest many were capable diggers, if not habitual ones. Digging can be seen to save lives in modern mammals too: Kangaroo rats for example (*Dipodomys*), living at nuclear bomb test sites in Nevada in the United States, can survive on the periphery of destruction thanks to their complex burrows and caches of food.*

Eating insects is also a good plan for survival. Being insectivorous usually goes hand-in-hand with being small-bodied, so the naturally smaller birds and mammals were once again sitting pretty.

There are many possible reasons why some made it through and flourished when others did not. One of them, as is always the case in the story of life on Earth, is the luck of the draw.

What happened to mammals right after the K–Pg mass extinction is currently poorly known. Part of the problem is a patchy fossil record globally. As is so often the case in mammal palaeontology, most of these early Palaeogene mammals are known only from a few scattered teeth.

Most of the prolific Cretaceous groups that survived didn't hang around for long. The gondwanatherians of Madagascar were wiped out and colonisation of the isolated island started from square one, with placental mammals as primary ingredients. This explains the unique creatures that inhabit it today: lemurs, tenrec, euplerid carnivores like the fossa and a host of other endemic species.† Around 90 per cent of species on the island are found nowhere else, but face an uncertain future due to habitat destruction.

Multituberculates made it across the boundary and did well for a time, but they soon began to disappear.[11] Their decline has been linked to the rise of rodents, but the picture of the interaction between these superficially similar groups is unclear. Resource competition may have been a factor, but changes in faunal composition – as in the case of

---

* They do not, obviously, survive direct detonation.[9] However, a report from 1965 lists 45 species of mammal living at the Nevada test sites, proving how resilient life can be in the face of destruction.[10]
† The aye-aye (*Daubentonia madagascariensis*) gets a lot of flak as the weirdest of them all, but the award for best-named has to go to the bastard big-footed mouse, *Macrotarsomys bastardi*.

tritylodontids before them – might have played a larger role. The new kinds of predators that emerged, including creodontan carnivores and birds of prey, wouldn't have helped. By the Late Eocene (the second main epoch of the Palaeogene), multituberculates were gone.

The Palaeogene was the preserve of the unfussy, not only in terms of diet but also in anatomy. The first mammals of the new age were all quite similar to one another at first. Many are so hard to tell apart that historically they were lumped in a single wastebasket group called 'Condylarthra'. If you added up all of the mammals that had existed to that point, took an average, then made it larger-bodied, you'd end up with a condylarth. They are neither this nor that, a melting pot of mammal blueprints that thrived in the earliest days of the Palaeogene. A bit sturdy, medium sized, not really specialising in anything in particular – except resilience of course. They are placental mammals, of that much we are certain, but their exact taxonomy and relationships are not well understood, partly because they are all so alike. Some of the major orders of modern mammals stepped forward from among them, but until more complete material is discovered it's hard to be sure who's who.

What we do know is that the earliest members of all of the modern groups proliferated in the first 10–20 million years of the Palaeogene. Marsupials and their kin had been doing quite well across the northern hemisphere at the end of the Cretaceous, but not so much afterwards. They radiated into new groups in North America after the impact, then journeyed down through South America to reach Patagonia. This southernmost country was still attached to a green and fertile Antarctica in the Palaeogene. Animals thrived on Antarctica, including for a while those Cretaceous southerners, the gondwanatherians.[12] Like multituberculates, the gondwanatherians persisted for a time after the K–Pg, but their last fossils are known from the Early Eocene of Patagonia, around 59 million years ago.

Marsupials, meanwhile, dispersed across the richly forested polar continent, reaching Australia before it finally broke away. This is why we find marsupials in Australia and South America today, lost from their original northern ranges. Around 35 million years ago the ice sheets began to spread on Antarctica, pushing the lost polar ecosystems to the fringes. There, they shrivelled and finally disappeared under the glacial press, leaving only their fossils behind.

Monotremes continued their long plod through time – if it ain't broke, don't fix it. A fossil from Patagonia named *Monotrematum* gives away the presence of their ancestors there around 61 million years ago, not long after the mass extinction. The fossil of *Obdurodon* shows us that by 28 million years ago there was a recognisably platypodian inhabitant in Australia, albeit retaining its ancestral molar teeth, which are lost in the modern species (*Obdurodon* means 'enduring teeth').

Among placental mammals, however, families rapidly diverged on almost every continent. The first afrotherians – the lineages including elephants, aardvarks and golden moles – emerged in Africa. Meanwhile in the north, Laurasiatherians were treading new ground: the ancestors of bats (Chiroptera), and whales and hoofed mammals (Cetartiodactyla) appeared. The carnivorans split up with the pangolins. Shrews knew a good thing when they saw it, and continued to live a similar lifestyle to many early mammals.

Meanwhile a third big branch of placental mammals was taking shape, the Euarchontoglires. These include rodents, the super-success story of the modern age. Near the end of the Palaeogene they ran in and took the seat from multituberculates in the game of evolutionary musical chairs. Lagomorpha, the rabbits and hares, sought an early divorce from the rodents, but kept some of the same habits. Similarly tree-shrews and colugos set off on their own spindly branch, out on a limb separated from a particularly unremarkable little band of tree-huggers called plesiadapiforms.

Around 58 to 55 million years ago, *Plesiadapis* and animals like it were among the first animals we might consider our closer kin. They are less than a metre long, and look squirrely, with long gnawing front teeth and a diastema. Their eyes face forwards, providing depth-perception for a life lived in the trees – although some researchers dispute this, and place them as ground-dwellers. Either way, looking at their skeleton, a number of characters finger them as belonging to the base of our primate branch.

The descendants of plesiadapiforms have been monkeying around ever since. They gave rise to lemurs, monkeys, primates – and of course one of their latest and most controversial of offshoots: humans.

◦ ◇ ◦

Recovery from the K–Pg in Scotland might have been slower thanks to the geological drama that took place there. Some of Scotland's most spectacular landscapes have their origin in the Palaeogene. As America and Europe broke apart, the Highlands and islands – *A' Ghàidhealtachd*, 'place of the Gaels' – were pushed upwards by the rising bile of lava. The volcanoes of Skye and the rest of the Inner Hebrides, Ardnamurchan and St Kilda, emptied their stomachs over what was left of the Cretaceous, in a similar fashion to the Siberian traps at the end of the Permian.

Recent research by geologists suggests the universe sent helpers to aid Skye's volcanic tantrum. Rocks on the island from around 60 million years ago contain minerals from asteroids.[13] They were only small impacts, but may have triggered some of the volcanic activity in the area. Life on Earth was coming back from the brink of global disaster, but in Scotland it probably remained a difficult place to survive at the start of the Palaeogene.

For the animals and plants that clung on, the climate was tropical – more like the coast of West Africa today. Global warming in the Palaeogene steadily turned up the heat until it peaked during what scientists call the PETM, Palaeocene-Eocene Thermal Maximum.* This increase in temperatures, by as much as 8° Celsius globally, had repercussions for the patterns of modern mammal evolution. Many emerging groups reduced in body size in response to the shifting floral composition and changes in climate.

Fossils from the Palaeogene in Scotland are almost non-existent, so understanding life there at the time is difficult. There are some beetle wing-cases from the Isle of Mull, but little else. The rocks that might have contained Palaeogene fossils were first weathered by natural processes, then scraped off by recent ice ages.

The deposits left in England, however, give us a glimpse of what populated that fresh post-K–Pg world in this tiny northern outpost. The tooth of an animal called *Arctocyonides* is the earliest vertebrate fossil from the Palaeogene in the British Isles.[14] It is around 57 million years old, and has been thrown in the condylarth wastebasket. It could

---

* The Palaeocene is the first part of the Palaeogene, the Eocene is the middle part, and the Oligocene is the last part.

belong to an early hoofed mammal, or maybe even a member of one of the mammal groups that didn't survive to the present, the carnivorous creodonts.

By the time we reach the middle Palaeogene the whole cast of characters are assembled in the rocks of southern England. Teeth and bones belonging to early members of the primates, rodents, rabbits, hoofed mammals and some slinky carnivorans have been found. A few metatherians are among them, hanging on to their last strongholds until the bitter end. Marsupials remain in Europe until the middle of the Miocene (around 14 million years ago) – the last of their remains have been found in Germany.

The oldest known mole, *Eotalpa* (the 'dawn mole') comes from the middle Palaeogene of England.[15] Rediscovering the lifestyle pioneered by docodonts over 100 million years before, it occupied a landscape covered in forest, sharing it with diversifying lineages of birds, lizards and amphibians.

These are mere teasers of the life that flourished globally as the world recovered from the impact. Recent glaciers may have removed some of Scotland's palaeontological past, but they also exposed the Mesozoic rocks of the Inner Hebrides. The ice carved out corries, sharpened some peaks and rubbed the others as smooth as cue tips. Pouring through the gaps between the Palaeogene volcanoes, it pulled up just enough of the basalt blanket to let us take a peek at the Jurassic ecosystem. What nature takes away with one hand, it gives with another.

Research into Scotland's ancient mammals has only just begun.[*] It is a small but significant moment in time written into the rock in script so tiny we need powerful synchrotron X-rays to read it. These bones are part of a global fossil dataset, and as we begin to add new discoveries and re-examine old ones, our picture of the changing world is continually updated. New visual analytical techniques bolster observations and give surprises. The crunching of calculations – thanks to computers, coding and a dense mesh of mathematical

---

[*] My and my co-author's first major paper on Waldman and Savage's Jurassic mammal skeletons came out earlier this year, and there are more to follow. You can finally meet these ancient inhabitants of our islands.

analyses – adds quantitative back-up to insightful observations. The chemistry of rocks provides our baking time, while isotopes and genetics add flavour to the meal.

Palaeontology, once a Euro-American pursuit led by white men with museums to fill and frontiers to plunder, is rightfully being prised back by researchers from the rest of the globe. Women are increasingly recognised for their contributions – historical and ongoing. Fossils are being repatriated.

But there is still much work to do to ensure those who study this science are as diverse as the life they research. We are moving away from science as the work of a few eccentric geniuses, into the age of collaboration. No ecosystem thrives without diversity. Our work is richer for it.

◇ ◇ ◇

And so our journey ends where most other mammal origin stories only just begin. Selfish creatures that we are, we like to hear stories about ourselves, but I won't indulge you any further.

One might argue that this whole book is the work of a navel-gazer, placing mammals on a storytelling pedestal over reptiles and amphibians. But that was not the intention. I hope it has given you a different perspective on our evolutionary history, one that is not overpowered by giant flashy reptiles. There are still many more stories to tell, but I can only fit so much of evolution's wonders into a single book.

From their origins in the Carboniferous, synapsids have made it through countless extinctions, and tried out every way to make a living. At times they've taken charge and blown the top off the definition of success – and weirdness. At others, they quietly innovated, as the world enlarged around them. I hope I've proven in these pages that it is not our recent branches, but the many other players in the synapsid game, who have pulled the best moves in the last 300 million years.

The journey of the most recent 'age of mammals' has been told countless times. It is a path that starts 66 million years ago in the Palaeogene and swiftly sets off in a thousand directions. Whichever trail you take, the destination is the present day.

You know the way home from here.

Triumph of the Little Guy

… for there is no folly of the beasts of the earth which is not infinitely outdone by the madness of men.

Herman Melville, *Moby Dick*

Around the world, young people took to the streets on 15 March 2019. They walked out of schools and colleges. They pulled their parents from their homes by the hand. Their faces were daubed in flowers – angiosperms once again at the heart of a revolution. Small fingers clutched banners splashed with scarlet fires, melting Earths and exclamation marks. In town and city centres they pooled, and their small voices swelled. They are the next generation, calling for their elders to take action in a climate emergency.

Anthropogenic climate change is reshaping our planet and human politics. I joined the young people in Oxford that day marching through the streets with the sun in our eyes, hoping that politicians and big business would not be similarly blinded to the crisis taking place around us. It was a joyful but melancholy protest. There was a promising future in those pleading faces, but the people most determined to fix the problem are the ones with the least political power to enact timely change. We just have to hope their parents and grandparents are listening.

How to deal with the inevitable consequences of climate change – rising sea levels, extreme weather, frequent wildfires, shifting baselines – is the most urgent concern for humankind in the next century and beyond. Our struggle is environmental, social, political and economic. Unlike every other living thing that has preceded us, we actually have a choice about how to survive, and the foresight and planning to pre-emptively make those changes.

For non-human animals, surviving mass extinction is an evolutionary lottery. Whether they'll have the right traits to see them through isn't something they can work towards, but a game of musical chairs. There are many more participants than there are seats to sit upon.

When it comes to death on a global scale, one thing we know is that it pays to be small. Whether that death falls from the sky, wells up from the heart of the Earth, or is the result of an industrious ape turning the atmospheric oven up to gas mark oblivion, the little guys are the ones who persist. Smallness is not the only criteria: the less fussy you are about food and habitat, the faster you breed, and the wider your geographic range, the better your chances. Survivors can't be choosers.

After the Permian, we see from the fossil record how a few creatures made the best of a bad situation. Among synapsids, time and again they've put forward a candidate to carry them through: *Lystrosaurus*, the piggy persister, virtually took over the world 252 million years ago. Small burrowing placental mammals were first off the starting block when the asteroid sent dinotopia sky-high at the end of the Cretaceous. Disaster taxa can be identified after every one of the world's mass-extinction events. Sustaining that success, well, that's the tricky part.

With so much changing around us, what does the study of evolution tell us about who might make it through our current extinction crisis? How will non-human mammals fare in the world we are creating?

◇ ◇ ◇

Humans are replicating many of the conditions of previous mass extinctions. The greenhouse effect caused by the release of carbon dioxide and methane into our atmosphere is raising the global temperature, just as the Siberian flood basalts did with their gases 252 million years ago. Our gobbling of habitat for building, agriculture and industry is not unlike the destruction wrought by the asteroid impact 66 million years ago and its aftermath for vegetation globally. Although these things have happened before on much bigger scales, they usually occur over a much longer time period, providing leeway for some organisms to adjust, and then they abate, providing time for recovery.

As our climate warms it is possible that mammals may become more active at night.[1] In arid and hot environments such as deserts, animals restrict their movements until after dark. For mammals this is to prevent overheating, hyperthermia, which is a lot more dangerous for endotherms than being too cold, hypothermia. Over half a million people die each year when their body temperature rises too far, experiencing organ failure and brain damage before succumbing. With the number of days on Earth over 35° Celsius (95° Fahrenheit) expected to triple in the coming decades, hyperthermia is a real danger not just to humans, but to all animals.

Studies have shown that many mammals (as well as some amphibians, fish and insects) have flexibility in their activity patterns. They change the times they forage or hunt to compensate for changes in season, daily temperature, drought and predation risk. This is known as temporal flexibility, and is relatively common. As we discovered in earlier chapters, most mammals today are active at dusk, dawn or during the night, and they retain the colour blindness of their nocturnal ancestors. This provides them with the flexibility to alter their activity in response to rising daytime temperatures by returning to the niche exploited by the earliest mammals in the Late Triassic.

As global temperatures rise we will also see an increase in the frequency of droughts. Animals will face challenges retaining moisture. Being nocturnal helps with this too, conserving water and energy that would otherwise be wasted on efforts to cool down.

For those animals that can't shift activity patterns, the best response is moving – either latitude or altitude. Studies have shown this is already taking place, with animals in high latitudes retreating pole-ward, while those elsewhere have started climbing.[2] This echoes the pattern seen after the Permian mass extinction, when fearsome temperatures at the equator made it uninhabitable for most organisms.

However moving is not a solution, only a response. Animals can't forward plan, so their movements are actually random. A paper published in 2013 estimated that their wanderings actually negatively impacted their chances of survival by almost 90 per cent, and would only benefit organisms in around 5 per cent of locations globally.[3] Looking back in time, as the last major glaciation ended in Europe around 11,000 years ago, populations of Arctic fox (*Alopex lagopus*) in the southern half of the continent failed to follow the retreating ice northwards; they just became extinct below a certain latitude.[4]

Most animals, including ourselves, will struggle in the hothouse of the future, but those with the largest geographical ranges have the most tickets to try and win the climate-change lottery.

Although relocating could help certain species – and may have saved some bacon in the past – there is one important difference in the modern day: the presence of humans. We place multiple barriers against animal dispersal, including infrastructure such as roads, urban environments, fences, canals and altered habitats (like fields). These prevent natural movement that could otherwise allow species to seek refuge. Research has shown that around half of species are currently on the move, and on land they are migrating at around 15 kilometres (10 miles) per year. There are few places where you can travel that distance without bumping into something human-made.

Size, as we discussed already, is an important factor in surviving mass extinction. The Lilliput effect means smaller animals survive better, and larger animals tend to become smaller. This is bad news for the charismatic megafauna so beloved of safaris, zoos and wildlife documentaries. The problems of heat and drought disproportionately affect larger animals due to their lower surface-to-volume ratio, hindering their ability to shed excess heat. Loss of habitat impacts their food sources and movements too, stacking up the challenges. A study published in 2020 looking at the distribution of mammals since the 1970s found that declines in range correlated with large body mass, increased air temperature, loss of habitat and high density of humans.[5] Conversely, a few small generalists with high reproduction rates had actually expanded their range. If I were you, I'd say goodbye to any wild animal bigger than a pig – zoos are likely to be the only refuges for them in the future we are creating.

Being unfussy has already proven an effective strategy for surviving with humans. Rats and mice, the vermin we fear and poison, love living alongside us because we provide such rich pickings. Leftover chips, a fruit bowl on the kitchen counter, bird seed in the garden, crops or even plastic cabling, you name it and they'll consume it. Certain species of fox, opossums and racoons have also discovered what we're good for, adapting to our noise and built environments in exchange for the perfectly good detritus we discard daily. These urban species thrive in higher numbers and have larger litters than their countrified counterparts.

Sea-level rises have already claimed mammal victims. Among them is the Australian rodent, the Bramble Cay melomys (*Melomys rubicola*), which in 2016 was given the unhappy title of the first mammal to become extinct due to anthropogenic climate change. Rising sea levels and severe weather destroyed its fragile coral atoll habitat, proving that being small might stack your chances, but it certainly doesn't make you immune to extinction.

You might have thought rising seas could benefit marine mammals, but ocean acidification and deoxygenation are having such detrimental impacts on the food chain that they far outweigh any advantages. In mass extinctions past, large marine predators have tended to fare poorly. A study in 2020 suggests that marine mammals are already facing an increase in the occurrence of infectious diseases, which could accelerate as extreme seasonal weather intensifies.[6]

Habitat destruction is the most insidious driver of extinction. This is not only true for anthropogenic extinction, but in the fossil record changes in flora have always had the most devastating repercussions for other organisms. Our use of land for agriculture is the major driver of biodiversity collapse, as we try to feed 7.8 billion people and more each year. Around 70 per cent of that agriculture is related to meat production, which also gobbles vast volumes of water and grain. This is part of the reason that becoming vegetarian or vegan is one of the most powerful ways an individual can act to save the world.

Around 60 per cent of all mammals alive today are kept by humans for food. There are over 1.4 billion cows, 1.8 billion sheep and goats and 980 million pigs,[7] which alongside humans account for 34 per cent of all mammals on Earth. Our pet cats total 600 million, our dogs 470 million. Wild mammals, meanwhile, now represent only 4 per cent of all mammals alive today.*

---

* Birds are doing a little better, with 30 per cent of them wild, 70 per cent livestock – almost all chickens.

This could be viewed as a phenomenal success story for the placental mammal groups we have selected and bred for meat and companionship, but it is a devastating picture for global diversity. The ancestors of cows and horses are extinct now, survived by mere tribute bands to their former glory.[*]

Livestock and humans comprise a paltry sliver of the stunning sum of 300 million years of synapsid evolution. With one in four mammal species at ongoing risk of extinction,[†] we are destroying everything our lineage has persisted for.

We are living in the 'age of humans', the Anthropocene. How ironic it is that its defining feature is the creation of a world that cannot sustain us.

◦ ◊ ◖

As I finish this book I'm sitting in the last days of the first full lockdown in Scotland. Our fieldwork on Skye was abandoned just over two months ago when the Covid-19 pandemic went from a threat on the horizon to a grim global reality. The human cost has been devastating, and continues, but the benefit for every other living thing – mammal or otherwise – has been almost entirely positive in the short term.

In the United Kingdom we were asked to stay home to save lives, preventing the spread of the virus. Restricted to a single hour of exercise outside per day, people walked their neighbourhoods, often for the first time in years. On these constitutionals we were soon struck by the burgeoning green spaces, which seemed twice as lush as before. Unmown roadside verges were quickly coated in spectacular wildflower blooms.

Darwin could have been describing a stroll mid-pandemic when he coined his famous paragraph, 'It is interesting to contemplate a tangled bank, clothed with many plants of many kinds, with birds singing on the bushes, with various insects flitting about ...'[8] Seeing the difference just a few weeks of unconstrained growth can make gives us a glimpse beneath the shifting baseline.

---

[*] For an amazing journey through the natural history of our domesticated species, read Richard Francis's *Domesticated: Evolution in a Man-Made World*.
[†] According to the International Union for Conservation of Nature (IUCN) Red List of Threatened Species.

With our streets empty of cars and pedestrians, wildlife has been taking the opportunity to explore. In Llandudno in Wales, feral goat herds ransacked the gardens of local residents, swaggering through the cul de sacs like thugs. Sika deer in Nara in Japan, usually fed by tourists, foraged in the suburbs. In Tel Aviv, jackals and wolves took themselves for walks in Hayarkon Park. Vehicle collisions have an insidious impact on wildlife populations, including mammals, with as many as a million wild animals usually killed on the highways of the United States every day.[9] Lockdown has probably saved many more animal lives than it has human ones.

In the first month of the pandemic, carbon dioxide emissions in China fell by a quarter.[10] For the first time in over 30 years residents in the Punjab in India could see the Himalayas, 200 kilometres (125 miles) distant, thanks to the reduction in air pollution. Pictures were shared on social media with the hashtag #GlobalHealing. People revelled in glimpses of a natural world their grandparents took for granted. This is the world we wish for, hidden under the one we never hoped to see.

Despite all this, the International Renewable Energy Agency (IRENA) estimates that by the end of 2020 annual CO_2 emissions will only have fallen by 6–8 per cent. Other predictions suggest the reduction will barely offset the steady rise,[11] and soon it will be business as usual once more. The 'green' start to 2020 is nothing compared to the scale of the problem. To make any difference we would need this lockdown to happen every year for the next 30 years.

We are now the asteroid. In this mass–extinction event, humanity is the lava spew. History is repeating itself, but unlike in the geological past there is an animal that can be proactive, not just reactive. As well as destroying, we also have the power to stop the extinction and restore our world.

Failing this, it's time to usher in a world of rats and cockroaches. We regard them with such disdain, but really we should envy their flexibility and resilience. Although many habitats will be lost, others will open up for new diversifications. What creatures will make use of Antarctica once humankind has finished stripping her of her glacial dignity?

We are a walking disaster taxon. Like other mammals before us we might turn to insects for nutrition, when global food and water shortages halt agriculture. Livestock and pets are unlikely to accompany us far into the future – although some will probably outlast us.

The mammals that survive this mass extinction will be the same as in the previous ones: small, burrowing, nocturnal, generalists. It's a way of life they've been practising for the last 210 million years. They're pretty good at it.

We face uncertain times. If there is one great consolation to be found in the fossil record, it's that life always survives somehow. While I have no doubt that synapsids will continue to run the race as they always have done, it probably won't be we humans who carry the baton.

Acknowledgements

Before anything else, I must acknowledge that many of the fossils described in this book were found on Indigenous lands, where researchers and collectors often took specimens without consultation or permission. This occurred prolifically into the early twentieth century owing to the colonial attitudes towards people outside Europe. Researchers, including myself, continue to benefit from these resources. Science is structured by the disproportionate legacy of white European thought and culture, inherently excluding people of colour and erasing their contributions. I want to pay my respects to the traditional stewards of the lands where these fossils come from, and acknowledge that they are the true custodians of their natural heritage.

There are so many people to thank for helping make this book come to life. Firstly my editors, Jim Martin, Anna MacDiarmid and Angelique Neumann, for taking me on in the first place, checking up on me during lockdown, and helping to make the book better with feedback and guidance, and to Catherine Best for her feedback and meticulous editing. April Neander used her skill bringing the past to life with her art, and I feel truly blessed she agreed to work with me.

I'm very grateful to all the people who were generous with their time by giving feedback on my text, including Ross Barnett, Julien Benoit, Neil Brocklehurst, Jonah Choiniere, Vincent Fernandez, David Ford, Nick Fraser, Christine Janis, Christian Kammerer, Susannah Lydon, Zhe-Xi Luo, Chris Manias, Julia Panciroli, Dugald Ross, Lara Sciscio and Michael Waldman. It was so helpful to get specific information from people who are experts in their field: Lara Sciscio gave me information on detrital zircons and the Karoo; Kimi Chapelle taught me about Karoo sauropods; Pia Viglietti told me about the experiences of Ione Rudner; Gwen Antell talked to me about decolonising science; Bernhard Zipfel was one of the first to speak openly to me about the legacy of Robert Broom; and Christa Kulijan gave me advice and information on how to approach writing about the unpleasant history of science and race.

Images in the colour plates were given for use by Julien Benoit, Matt Humpage, Zhe-Xi Luo, the Oxford University Museum of Natural History, the Field Museum of Natural History in Chicago and the Royal

Tyrrell Museum of Palaeontology in Drumheller. Photos of Zofia Kielan-Jaworowska are courtesy of Jolanta Kobylinska and Magdalena Borsuk-Bialynicka, who provided them from the incredible archive of images at the Institute of Palaeobiology in Poland, which are currently being digitised. Thank you also to the talented writers and poets who gave me permission to reproduce their work: Màiri Anna NicUalraig (Mary Ann Kennedy), Justin Sales and Fiona Ritchie Walker. I can't wait to read more!

On a more personal note, I'm so grateful to Roger Benson, Stig Walsh, Richard Butler and Jonah Choiniere, who are such good mentors, great company, and thankfully didn't take offence at my descriptions of them. A special thanks to Roger Benson, who has been unflinchingly supportive (and constructively critical) of all my work, and who accommodated my mini 'sabbatical' to write this book, which thanks to the coronavirus lockdown ballooned into a five-month apocalyptic writing retreat. Thanks also to Michael Waldman for all the time he spent telling me about his life, research and the history of the Skye specimens – without his discoveries I wouldn't have ended up where I am. Many thanks to my academic mentors Zhe-Xi Luo, Nicholas Fraser and Stig Walsh for supporting me through everything; without them I'd never have made it through my PhD or had the mental space to start writing this book at the same time.

Although they didn't know it, my friends and followers online (especially on Twitter) have been consistently refreshing my enthusiasm for writing, by being excited about the book and its topic – thank you! There are many, many people who played huge roles in the scientific discovery and research described in this book, but who couldn't be mentioned due to space constraints and the purposes of keeping the narrative going. These include John Hudson, geologist and expert on the Hebrides, and my fellow mammals workers too numerous to list. You are all such incredibly important people, and I hope you'll forgive me for the inevitable errors and omissions.

Finally I need to give a massive loving hug to my mum and my husband for accepting that I was often glued to my laptop and not paying attention. I wrote this thinking always of my dad, who I miss every day and wish was here to see this book published; he'd be so proud. And love to my brother and sisters and the rest of my family for always being so supportive of me and my sometimes unusual interests. I've been boring you all with natural history talk for years, and I have every intention of continuing to bore you for years to come.

Finally, thank you, reader, for picking up this book. I hope you enjoy it.

Notes

Chapter 1

1. Forbes, A. R. 1923. *Place names of Skye and Adjacent Islands: With Lore: Mythical, Traditional, and Historical.* Alexander Gardner Ltd, Paisley: p. 333.

Chapter 2

1. Buckland, W. 1824. Notice on the Megalosaurus or the Great Fossil Lizard of Stonesfield. *Geological Society London* 2: 390–396.
2. Ibid., p. 391.
3. Hakewell, H. Esq. 1822. Notice on the Stonesfield Slate pits by MGS. In Howlett, E. A., Kennedy, W. J., Powell, H. P. & Torrens, H. S. 2017. New light on the history of Megalosaurus, the great lizard of Stonesfield. *Archives of Natural History* 44: 82–102.
4. Buckland, F. T. 1858. Memoir of the Very Rev. William Buckland, D.D., F.R.S., Dean of Westminster. In Buckland, W. *Geology and mineralogy considered with reference to natural theology.* Routledge & Company, London.
5. Charles Lyell to Lyell Senior, 24 August 1828. In Rudwick, M. 2008. *Worlds Before Adam: The Reconstruction of Geohistory in the Age of Reform.* University of Chicago Press, Chicago: pp. 270–1.
6. Charles Lyell to Lyell Senior, 14 November 1827. In ibid., p. 248.
7. Owen, R. 1834. On the generation of the marsupial animals, with a description of the impregnated uterus of the Kangaroo. *Philosophical Transactions of the Royal Society of London* 124: 333–364.
8. Owen, R. 1871. Monograph of the Fossil Mammalia of the Mesozoic Formations. *Monographs of the Palaeontographical Society* 24: 1–115, 111.
9. Ibid., 114.
10. Ibid., 112.
11. Osborn, H. F. 1887. On the structure and classification of the Mesozoic Mammalia. *Proceedings of the Academy of Sciences of Philadelphia*: 282–292, 287.
12. Ibid., 291.

Chapter 3

1. Janis, C. M. & Keller, J. C. 2001. Modes of ventilation in early tetrapods: Costal aspiration as a key feature of amniotes. *Acta Palaeontologica Polonica* 46: 137–170.
2. Mann, A., Gee, B. M., Pardo, J. D., Marjanović, D., Adams, G. R., Calthorpe, A. S., Maddin, H. C. & Anderson, J. S. 2020. Reassessment of historic 'microsaurs' from Joggins, Nova Scotia, reveals hidden diversity in the earliest amniote ecosystem. *Papers in Palaeontology* 6: 605–625.

Chapter 4

1. A study found that over 75 per cent of history books published in America were written by men, and of biographical books, 71 per cent were about men. Kahn, A. & Onion, R. 6 January 2016. Is History Written About Men, by Men? *The State*, www.thestate.com

2. Imogen Robertson, Chair of the Historical Writers' Association in the United Kingdom, told an interviewer: 'The gender bias is horribly clear when you look at what women are supposed to write about: men write about grand expansive histories, the thesis-driven revisionist world views, and they mostly have the Second World War to themselves.' Rottner, T., 5 March 2016. Is History Written by Men? *The Bubble*, www.thebubble.org.uk

3. Based on a translation featured in a Wikipedia entry for Roderick Impey Murchison.

4. Duncan, H. 1831. An account of the tracks and footmarks of animals found impressed on sandstone in the quarry of Corncockle Muir, in Dumfriesshire. *Proceedings of the Royal Society of Edinburgh* 11: 194–209.

5. Grierson, J. 1828. On Footsteps before the Flood, in a specimen of red sandstone. *Edinburgh Journal of Science* 8: 130–134.

6. Duncan, H. 1831. An account of the tracks and footmarks of animals found impressed on sandstone in the quarry of Corncockle Muir, in Dumfriesshire. *Proceedings of the Royal Society of Edinburgh* 11: 194–209.

7. Ibid.

8. Cope, E. D. 1878. Descriptions of Extinct Batrachia and Reptilia from the Permian Formation of Texas. *Proceedings of the American Philosophical Society* 17: 505–530.

9. Mayor, A. 2005. *Fossil Legends of the First Americans*. Princeton University Press, Princeton: pp. 195–8.

10. Cope, E. D. 1886. The long-spined Theromorpha of the Permian epoch. *The American Naturalist* 20: 544–545.

11. Ibid.

12. Haack, S. C. 1986. A thermal model of the sailback pelycosaur. *Paleobiology* 12: 450–458.

13. Bennett, S. C. 1996. Aerodynamics and thermoregulatory function of the dorsal sail of *Edaphosaurus. Paleobiology* 22: 496–506.

14. Huttenlocker, A. K., Mazierski, D. & Reisz, R. R. 2011. Comparative osteohistology of hyperelongate neural spines in the Edaphosauridae (Amniota: Synapsida). *Palaeontology* 54: 573–590.

15. Bailey, J. B. 1997. Neural spine elongation in dinosaurs: sailbacks or buffalo-backs? *Journal of Paleontology* 71: 1124–1146.

16. Rega, E. A., Noriega, K., Sumida, S. S., Huttenlocker, A., Lee, A. & Kennedy, B. 2012. Healed fractures in the neural spines of an associated skeleton of *Dimetrodon*: implications for dorsal sail morphology and function. *Fieldiana Life and Earth Sciences* 5: 104–111.

17. Darwin, C. 1859. *On the Origin of Species* (1st edition). John Murray, London: p. 88.

18. Cooper, N., Bond, A. L., Davis, J. L., Portela Miguez, R., Tomsett, L. & Helgen, K. M. 2019. Sex biases in bird and mammal natural history collections. *Proceedings of the Royal Society B* 286: 20192025.

19. Cope, E. D. 1880. Second contribution to the history of the Vertebrata of the Permian formation of Texas. *Proceedings of the American Philosophical Society* 19: 38–58.

20. Clauss, M., Frey, R., Kiefer, B., Lechner-Doll, M., Loehlein, W., Polster, C., Rössner, G. E. & Streich, W. J. 2003. The maximum attainable body size of herbivorous mammals: morphophysiological constraints on foregut, and adaptations of hindgut fermenters. *Oecologia* 136: 14–27.

Chapter 5

1. Barghusen, H. R. 1975. A Review of Fighting Adaptations in Dinocephalians (Reptilia, Therapsida). *Paleobiology* 1: 295–311.

2. Randau, M., Carbone, C. & Turvey, S.T. 2013. Canine evolution in sabretoothed carnivores: natural selection or sexual selection? *PLOS One* 8: e72868.

3. Cisneros, J. C., Abdala, F., Rubidge, B. S., Dentzien-Dias, P. C. & de Oliveira Bueno, A. 2011. Dental occlusion in a 260-million-year-old therapsid with saber canines from the Permian of Brazil. *Science* 331: 1603–1605.

4. Cisneros, J. C., Abdala, F., Jashashvili, T., de Oliveira Bueno, A. & Dentzien-Dias, P. 2015. *Tiarajudens eccentricus* and *Anomocephalus africanus*, two bizarre anomodonts (Synapsida, Therapsida) with dental occlusion from the Permian of Gondwana. *Royal Society Open Science* 2: 150090.

5. Froebisch, J. & Reisz, R. R. 2011. The postcranial anatomy of *Suminia getmanovi* (Synapsida: Anomodontia), the earliest known arboreal tetrapod. *Zoological Journal of the Linnean Society* 162: 661–698.

6. Ford, D. P. & Benson, R. B. 2020. The phylogeny of early amniotes and the affinities of Parareptilia and Varanopidae. *Nature Ecology & Evolution* 4: 57–65.

7. Chinsamy-Turan, A. 2012. *Forerunners of Mammals*. Indiana University Press, Bloomington: p. 281.

8. Ireland, A., Maden-Wilkinson, T., McPhee, J., Cooke, K., Narici, M., Degens, H. & Rittweger, J. 2013. Upper limb muscle-bone asymmetries and bone adaptation in elite youth tennis players. *Medicine and Science in Sports and Exercise* 45.

9. Macintosh, A. A., Pinhasi, R. & Stock, J. T. 2017. Prehistoric women's manual labor exceeded that of athletes through the first 5500 years of farming in Central Europe. *Science Advances* 3: eaao3893.

10. Montes, L., Le Roy, N., Perret, M., De Buffrenil, V., Castanet, J. & Cubo, J., 2007. Relationships between bone growth rate, body mass and resting metabolic rate in growing amniotes: a phylogenetic approach. *Biological Journal of the Linnean Society* 92: 63–76.

11. Rey, K., Amiot, R., Fourel, F., Abdala, F., Fluteau, F., Jalil, N. E., Liu, J., Rubidge, B. S., Smith, R. M., Steyer, J. S. & Viglietti, P. A. 2017. Oxygen

isotopes suggest elevated thermometabolism within multiple Permo-Triassic therapsid clades. *Elife* 6: e28589.

12. Betts, H. C., Puttick, M. N., Clark, J. W., Williams, T. A., Donoghue, P. C. & Pisani, D. 2018. Integrated genomic and fossil evidence illuminates life's early evolution and eukaryote origin. *Nature Ecology & Evolution* 2: 1556–1562.

Chapter 6

1. Quote from *The Economist, Farish Jenkins.* 17 November 2012: 98.
2. Figure from Clark, T. 21 August 2019. The Rock topped Forbes' list of the highest-paid actors in the world, which also includes five from the Marvel Cinematic Universe. *Business Insider.*
3. Stanley, S. M. 2016. Estimates of the magnitudes of major marine mass extinctions in earth history. *Proceedings of the National Academy of Sciences* 113: 6325–E6334.
4. Masaitis, V. L. 1983. Permian and Triassic volcanism of Siberia. *Zapiski VMO* 4: 412–425 (in Russian).
5. McElwain, J. C. 2018. Paleobotany and global change: Important lessons for species to biomes from vegetation responses to past global change. *Annual Review of Plant Biology* 69: 761–787.
6. Stanley, S. M. 2016. Estimates of the magnitudes of major marine mass extinctions in earth history. *Proceedings of the National Academy of Sciences* 113: 6325–6334.
7. Sciscio, L., de Kock, M., Bordy, E. & Knoll, F. 2017. Magnetostratigraphy across the Triassic–Jurassic boundary in the main Karoo Basin. *Gondwana Research* 51: 177–192.

Chapter 7

1. Jones, K. E., Angielczyk, K. D. & Pierce, S. E. 2019. Stepwise shifts underlie evolutionary trends in morphological complexity of the mammalian vertebral column. *Nature Communications* 10: 5071.
2. Jones, K. E., Angielczyk, K. D., Polly, P. D., Head, J. J., Fernandez, V., Lungmus, J. K., Tulga, S. & Pierce, S. E. 2018. Fossils reveal the complex evolutionary history of the mammalian regionalized spine. *Science* 361: 1249–1252.
3. Findlay, G. 1972. *Dr. Robert Broom, F.R.S.: Palaeontologist and Physician, 1866–1951: A Biography, Appreciation and Bibliography.* A. A. Balkema, Amsterdam: p. 25.
4. See Chapter 2 of Kuljian, C. 2016. *Darwin's Hunch: Science Race and the Search for Human Origins.* Jacana Media, Johannesburg.
5. Letter from Henry Fairfield Osborne, p. 41 in Findlay, G. 1972. *Dr. Robert Broom, F.R.S.: Palaeontologist and Physician, 1866–1951: A Biography, Appreciation and Bibliography.* A. A. Balkema, Amsterdam.

6. As pictured in Figure 15 in Crompton, A. W. 1968. In Search of the 'Insignificant'. *Discovery: Magazine of the Peabody Museum of Natural History, Yale University.* 3: 23–32.

7. As reported in *The Star,* 28 February 1952. In Kuljian, C. 2016. *Darwin's Hunch: Science Race and the Search for Human Origins.* Jacana Media, Johannesburg.

8. Story of her discovery as told by South African palaeontologist Dr Pia Viglietti, after meeting Ione Rudner.

9. Wallace, D. R. 2004. *Beasts of Eden: Walking Whales, Dawn Horses, and Other Enigmas of Mammal Evolution.* University of California Press, Berkeley: p. 129.

10. Angielczyk, K. D. & Schmitz, L. 2014. Nocturnality in synapsids predates the origin of mammals by over 100 million years. *Proceedings of the Royal Society B: Biological Sciences* 281: 20141642.

11. Lautenschlager, S., Gill, P. G., Luo, Z.-X., Fagan, M. J. & Rayfield, E. J. 2018. The role of miniaturization in the evolution of the mammalian jaw and middle ear. *Nature* 561, 533–537.

12. Warren, W. C., Hillier, L. W., Graves, J. A. M., Birney, E., Ponting, C. P., Grützner, F., Belov, K., Miller, W., Clarke, L., Chinwalla, A. T. & Yang, S. P. 2008. Genome analysis of the platypus reveals unique signatures of evolution. *Nature* 453: 175–183.

13. Hurum, J. H., Luo, Z.-X. & Kielan-Jaworowska, Z. 2006. Were mammals originally venomous? *Acta Palaeontologica Polonica* 51: 1–11.

14. Newham, E., Gill, P. G., Brewer, P., Benton, M. J., Fernandez, V., Gostling, N. J., Haberthür, D., Jernvall, J., Kankaanpää, T., Kallonen, A., Navarro, C., Pacureanu, A., Richards, K., Robson Brown, K., Schneider, P., Suhonen, H., Tafforeau, P., Williams, K. A., Zeller-Plumhoff, B. & Corfe, I. J. 2020. Reptile-like physiology in Early Jurassic stem-mammals. *Nature Communications,* 11: 1–13.

15. Blackburn, T. I. M. & Gaston, K. 1998. The distribution of mammal body masses. *Diversity and Distributions* 4: 121–133.

Chapter 8

1. Fernandez, V., Abdala, F., Carlson, K. J., Rubidge, B. S., Yates, A. & Tafforeau, P. 2013. Synchrotron reveals Early Triassic odd couple: injured amphibian and aestivating therapsid share burrow. *PLOS One* 8: e64978.

2. Miller, H. 1858. *The Cruise of the Betsey, Or, A Summer Ramble Among the Fossiliferous Deposits of the Hebrides: With Rambles of a Geologist, Or, Ten Thousand Miles Over the Fossiliferous Deposits of Scotland.* Constable & Co, Edinburgh.

3. Described to me by Nick Fraser, National Museums Scotland, 14 Dec 2017.

4. Told to me by Rachel Wood, University of Edinburgh, in 2016.

5. Story as recounted to me by Michael Waldman, summer 2017.

6. Wind, J. 1984. Computerized X-ray tomography of fossil hominid skulls. *American Journal of Physical Anthropology* 63: 265–282.

7. Conroy, G. C. & Vannier, M. W. 1984. Noninvasive three-dimensional computer imaging of matrix-filled fossil skulls by high-resolution computed tomography. *Science* 226: 1236–1239.

8. From the history pages of the Oxford University Museum of Natural History's website [retrieved April 2020].

9. Sollas, W. J. 1903. A method for the investigation of fossils by serial sections. *Philosophical Transactions of the Royal Society of London B: Biological Sciences* 196: 257–263, 262.

10. Benoit, J., Manger, P. R. & Rubidge, B. S. 2016. Palaeoneurological clues to the evolution of defining mammalian soft tissue traits. *Scientific reports* 6: 25604.

11. Oftedal, O. T. 2012. The evolution of milk secretion and its ancient origins. *Animal: an international journal of animal bioscience* 6: 355–368.

12. Zhou, C.-F., Bhullar, B.-A., Neander, A., Martin, T., Luo, Z.-X. 2019. New Jurassic mammaliaform sheds light on early evolution of mammal-like hyoid bones. *Science* 365: 276–279.

Chapter 9

1. Luo, Z.-X., Meng, Q.-J., Ji, Q., Liu, D., Zhang, Y.-G. & Neander, A.I. 2015. Evolutionary development in basal mammaliaforms as revealed by a docodontan. *Science* 347: 760–764.

2. Ji, Q., Luo, Z.-X., Yuan, C.-X. & Tabrum, A.R. 2006. A swimming mammaliaform from the Middle Jurassic and ecomorphological diversification of early mammals. *Science* 311: 1123–1127.

3. Meng, J., Hu, Y., Wang, Y., Wang, X. & Li, C., 2006. A Mesozoic gliding mammal from northeastern China. *Nature* 444: 889–893.

4. Luo, Z.-X. 2011. Developmental Patterns In Mesozoic Evolution of Mammal Ears. *Annual Reviews of Ecology, Evolution, and Systematics* 42: 355–380.

5. Luo, Z.-X., Schultz, J. A. & Ekdale, E. G. 2016. Evolution of the middle and inner ears of mammaliaforms: the approach to mammals. In Clack J. A., Fay, R. R. & Popper, A. A. (eds). *Evolution of the Vertebrate Ear: Evidence from the Fossil Record*. Springer Handbook of Auditory Research 59: pp. 139–74.

6. Heffner, H. E. & Heffner, R. S. 2018. The evolution of mammalian hearing. *AIP Conference Proceedings* 1965: 130001.

Chapter 10

1. Kielan-Jaworowska, Z. 2013. *In Pursuit of Early Mammals*. Indiana University Press, Bloomington: p. 74.

2. Ibid.

3. Henkel, S. 1966. Methoden zur Prospektion und Gewinnung kleiner Wirbeltierfossilien. *Neues Jahrbuch für Geologie und Paläontologie*: 178–184.

4. Cromie, W. J. 24 May 2001. Oldest Mammal is found: Origin of mammals is pushed back to 195 million years. *The Harvard Gazette*.

5. Kühne, W. G. 1956. *The Liassic Therapsid Oligokyphus*. British Museum, London.

6. Panciroli, E., Walsh, S., Fraser, N. C., Brusatte, S. L. & Corfe, I. 2017. A reassessment of the postcanine dentition and systematics of the tritylodontid

Stereognathus (Cynodontia, Tritylodontidae, Mammaliamorpha), from the Middle Jurassic of the United Kingdom. *Journal of Vertebrate Paleontology* 37, 373–86.

7. Hoffman, E. A. & Rowe, T. B. 2018. Jurassic stem-mammal perinates and the origin of mammalian reproduction and growth. *Nature* 561: 104–108.

8. 20 March 2019. Super bloom tourists cause small town 'safety crisis'. *BBC News*.

9. Fu, Q., Diez, J. B., Pole, M., Ávila, M. G., Liu, Z. J., Chu, H., Hou, Y., Yin, P., Zhang, G. Q., Du, K. & Wang, X. 2018. An unexpected noncarpellate epigynous flower from the Jurassic of China. *Elife*: e38827.

10. Smith, S. A., Beaulieu, J. M. & Donoghue, M. J. 2010. An uncorrelated relaxed-clock analysis suggests an earlier origin for flowering plants. *Proceedings of the National Academy of Sciences* 107: 5897–5902.

11. Barba-Montoya, J., dos Reis, M., Schneider, H., Donoghue, P. C. & Yang, Z. 2018. Constraining uncertainty in the timescale of angiosperm evolution and the veracity of a Cretaceous Terrestrial Revolution. *New Phytologist* 218: 819–834.

12. Hochuli, P. A. & Feist-Burkhardt, S. 2013. Angiosperm-like pollen and Afropollis from the Middle Triassic (Anisian) of the Germanic Basin (northern Switzerland). *Frontiers in Plant Science* 4: 344.

13. In Darwin, C. 1859. *On the Origin of Species* (1st edition). John Murray, London.

14. Friis, E. M., Pedersen, K. R. & Crane, P. R. 2001. Fossil evidence of water lilies (Nymphaeales) in the Early Cretaceous. *Nature* 410: 357–360.

15. Gomez, B., Daviero-Gomez, V., Coiffard, C., Martín-Closas, C. & Dilcher, D. L. 2015. *Montsechia*, an ancient aquatic angiosperm. *Proceedings of the National Academy of Sciences* 112: 10985–10988.

16. Sun, G., Dilcher, D. L., Wang, H. & Chen, Z. 2011. A eudicot from the Early Cretaceous of China. *Nature* 471: 625–628.

17. Field, D. J., Benito, J., Chen, A., Jagt, J. W. & Ksepka, D. T. 2020. Late Cretaceous neornithine from Europe illuminates the origins of crown birds. *Nature* 579: 397–401.

18. Kielan-Jaworowska, Z. 2005. *Zofia Kielan-Jaworowska: An Autobiography.* Unpublished: p. 3.

19. Wituska, K. & Tomaszewski, I. 2006. *Inside a Gestapo Prison: The Letters of Krystyna Wituska, 1942–1944.* Wayne State University Press, Detroit.

20. Buffetaut, E., & Le Loeuff, J. 1994. The discovery of dinosaur eggshells in nineteenth-century France. In Carpenter, K., Hirsch, K., & Horner, J. (eds). *Dinosaur Eggs and Babies*. Cambridge University Press, New York: pp. 31–4.

21. Gregory, W. K. 1927. Mongolian Mammals of the 'Age of Reptiles'. *The Scientific Monthly* 24: 225–235.

22. Kielan-Jaworowska, Z. 2005 *Zofia Kielan-Jaworowska: An Autobiography.* Unpublished: p. 18.

23. Kielan-Jaworowska, Z., Presley, R. & Poplin, C. 1986. The cranial vascular system in taeniolabidoid multituberculate mammals. *Philosophical Transactions of the Royal Society of London B: Biological Sciences* 313: 525–602.

24. Cifelli, R. L. & Fostowicz − Frelik, Ł. 2016. Legacy of the Gobi Desert: Papers in Memory of Zofia Kielan-Jaworowska. *Acta Palaeontologica Polonica* 67: 13.

25. Boyce, C. K., Brodribb, T. J., Feild, T. S. & Zwieniecki, M. A. 2009. Angiosperm leaf vein evolution was physiologically and environmentally transformative. *Proceedings of the Royal Society B: Biological Sciences* 276: 1771–1776.

Chapter 11

1. Longrich, N. R., Tokaryk, T. & Field, D. J. 2011. Mass extinction of birds at the Cretaceous–Paleogene (K–Pg) boundary. *Proceedings of the National Academy of Sciences* 108: 15253–15257.
2. Longrich, N. R., Bhullar, B. A. S. & Gauthier, J. A. 2012. Mass extinction of lizards and snakes at the Cretaceous–Paleogene boundary. *Proceedings of the National Academy of Sciences* 109: 21396–21401.
3. Vajda, V. & McLoughlin, S. 2004. Fungal proliferation at the Cretaceous–Tertiary boundary. *Science* 303: 1489.
4. Rehan, S. M., Leys, R. & Schwarz, M. P. 2013. First evidence for a massive extinction event affecting bees close to the KT boundary. *PLOS One* 8: e76683.
5. Donovan, M. P., Iglesias, A., Wilf, P., Labandeira, C. C. & Cúneo, N. R. 2016. Rapid recovery of Patagonian plant–insect associations after the end-Cretaceous extinction. *Nature Ecology & Evolution* 1: 1–5.
6. Slater, G. J. 2013. Phylogenetic evidence for a shift in the mode of mammalian body size evolution at the Cretaceous–Palaeogene boundary. *Methods in Ecology and Evolution* 4: 734–744.
7. Grossnickle, D. M., Smith, S. M. & Wilson, G. P. 2019. Untangling the multiple ecological radiations of early mammals. *Trends in Ecology and Evolution* 34: 936–949.
8. Benson, R. B. J., Hunt, G., Carrano, M. T. & Campione, N. 2018. Cope's rule and the adaptive landscape of dinosaur body size evolution. *Palaeontology* 61: 13–48.
9. Anderson, A. O. & Allred, D. M. 1964. Kangaroo rat burrows at the Nevada Test Site. *The Great Basin Naturalist* 24: 93–101.
10. Jorgensen, C. D. & Hayward, C. L. 1965. Mammals of the Nevada Test Site. *Brigham Young University Science Bulletin, Biological Series* 6: Article 1.
11. Wilson, G. P., Evans, A. R., Corfe, I. J., Smits, P. D., Fortelius, M. & Jernvall, J. 2012. Adaptive radiation of multituberculate mammals before the extinction of dinosaurs. *Nature* 483: 457–460.
12. Goin, F. J., Reguero, M. A., Pascual, R., von Koenigswald, W., Woodburne, M. O., Case, J. A., Marenssi, S. A., Vieytes, C. & Vizcaíno, S. F. 2006. First gondwanatherian mammal from Antarctica. *Geological Society, London, Special Publications* 258: 135–144.
13. 20 November 2018. Meteorite hunters dig up 60 million-year-old site in Skye. *BBC News*.
14. Hooker, J. J. & Millbank, C. 2001. A Cernaysian mammal from the Upnor Formation (Late Palaeocene, Herne Bay, UK) and its implications for correlation. *Proceedings of the Geologists' Association* 112: 331–338.

15. Hooker, J. J. 2016. Skeletal adaptations and phylogeny of the oldest mole *Eotalpa* (Talpidae, Lipotyphla, Mammalia) from the UK Eocene: the beginning of fossoriality in moles. *Palaeontology* 59: 195–216.

Epilogue

1. Levy, O., Dayan, T., Porter, W. P. & Kronfeld-Schor, N. 2019. Time and ecological resilience: can diurnal animals compensate for climate change by shifting to nocturnal activity? *Ecological Monographs* 89: e01334.

2. Pecl, G. T., Araújo, M. B., Bell, J. D., Blanchard, J., Bonebrake, T. C., Chen, I. C., Clark, T. D., Colwell, R. K., Danielsen, F., Evengård, B. & Falconi, L. 2017. Biodiversity redistribution under climate change: Impacts on ecosystems and human well-being. *Science* 355: eaai9214.

3. Buckley, L. B., Tewksbury, J. J. & Deutsch, C. A. 2013. Can terrestrial ectotherms escape the heat of climate change by moving? *Proceedings of the Royal Society B: Biological Sciences* 280: 20131149.

4. Dalén, L., Nyström, V., Valdiosera, C., Germonpré, M., Sablin, M., Turner, E., Angerbjörn, A., Arsuaga, J. L. & Götherström, A. 2007. Ancient DNA reveals lack of postglacial habitat tracking in the arctic fox. *Proceedings of the National Academy of Sciences* 104: 6726–6729.

5. Pacifici, M., Rondinini, C., Rhodes, J. R., Burbidge, A. A., Cristiano, A., Watson, J. E., Woinarski, J. C. & Di Marco, M. 2020. Global correlates of range contractions and expansions in terrestrial mammals. *Nature Communications* 11: 1–9.

6. Sanderson, C. E. & Alexander, K.A. Unchartered waters: Climate change likely to intensify infectious disease outbreaks causing mass mortality events in marine mammals. *Global Change Biology* 26: 4284–4301.

7. Robinson, T. P., Wint, G. W., Conchedda, G., Van Boeckel, T. P., Ercoli, V., Palamara, E., Cinardi, G., D'Aietti, L., Hay, S. I. & Gilbert, M. 2014. Mapping the global distribution of livestock. *PLOS One*, 9: e96084.

8. Darwin, C. 1859. *On the Origin of Species* (1st edition). John Murray, London.

9. Seiler, A. and Helldin, J.O. 2006. Mortality in wildlife due to transportation. In *The Ecology of Transportation: Managing Mobility for the Environment*. Davenport, J. and Davenport, Julia L. (eds). Springer, Dordrecht: pp. 165–89.

10. Watts, J. & Kommenda, N. 23 March 2020. Coronavirus pandemic leading to huge drop in air pollution. *Guardian*.

11. According to Peduzzi, P. 11 May 2020. Record global carbon dioxide concentrations despite COVID-19 crisis. *United Nations Environment Programme* [online: retrieved May 2020].

Bibliography

Chapter 1

Panciroli, E., Benson, R. B. J., Walsh, S., Butler, R. J., Castro, T. A., Jones, M. E. & Evans, S. E. 2020. Diverse vertebrate assemblage of the Kilmaluag Formation (Bathonian, Middle Jurassic) of Skye, Scotland. *Earth and Environmental Transactions of the Royal Society of Edinburgh* 111: 135–156.

Stephenson, D. & Merritt, J. 2006. *Skye: a landscape fashioned by geology.* Scottish Natural Heritage, Edinburgh.

White, S. & Ross, D. 2019. *Jurassic Skye: Dinosaurs and Other Jurassic Animals of the Isle of Skye.* NatureBureau, Newbury.

Chapter 2

Buckland, W. 1824. Notice on the Megalosaurus or the Great Fossil Lizard of Stonesfield. *Geological Society London* 2: 390–396.

Mayor, A. 2005. *Fossil Legends of the First Americans.* Princeton University Press, Princeton.

Moyal, A. 2001. *Platypus: The Extraordinary Story of How a Curious Creature Baffled the World.* Smithsonian Books, Washington DC.

Osborn, H. F. 1887. On the structure and classification of the Mesozoic Mammalia. *Proceedings of the Academy of Sciences of Philadelphia*: 282–292.

Owen, R. 1871. Monograph of the Fossil Mammalia of the Mesozoic Formations. *Monographs of the Palaeontographical Society* 24: 1–115.

Rudwick, M. 2008. *Worlds Before Adam: The Reconstruction of Geohistory in the Age of Reform.* University of Chicago Press, Chicago.

Schiebinger, L. 1993. Why mammals are called mammals: gender politics in eighteenth-century natural history. *The American Historical Review* 98: 382–411.

Chapter 3

Angielczyk, K. D. 2009. *Dimetrodon* is Not a Dinosaur: Using Tree Thinking to Understand the Ancient Relatives of Mammals and Their Evolution. *Evolution: Education and Outreach* 2: 257–271.

Beerling, D. 2017. *The Emerald Planet.* Oxford University Press, Oxford.

Clack, J. A. 2002. *Gaining Ground: The Origin and Evolution of Tetrapods.* Indiana University Press, Bloomington.

Shubin, N. 2009. *Your Inner Fish: The amazing discovery of our 375-million-year-old ancestor.* Penguin, London.

Sues, H.-D. 2019. *The Rise of Reptiles: 320 Million Years of Evolution*. Johns Hopkins University Press, Baltimore.

Chapter 4

Cooper, N., Bond, A. L., Davis, J. L., Portela Miguez, R., Tomsett, L. and Helgen, K. M. 2019. Sex biases in bird and mammal natural history collections. *Proceedings of the Royal Society B* 286: 20192025.

Chapter 5

Chinsamy-Turan, A. 2012. *Forerunners of Mammals*. Indiana University Press, Bloomington.

Kemp, T. 2005. *Origin and Evolution of Mammals*. Oxford University Press, Oxford.

Chapter 6

Ezcurra, M. D., Jones, A. S., Gentil, A. R. & Butler, R. J. 2020. Early Archosauromorphs: The Crocodile and Dinosaur Precursors. In *Encyclopedia of Geology (2nd edition)*. Elsevier, Amsterdam.

Hallam, T. 2005. *Catastrophes and Lesser Calamities: The Causes of Mass Extinctions*. Oxford University Press, Oxford.

Kolbert, E. 2015. *The Sixth Extinction: An Unnatural History*. Bloomsbury, London.

Stanley, S. M. 2016. Estimates of the magnitudes of major marine mass extinctions in earth history. *Proceedings of the National Academy of Sciences* 113: 6325–6334.

Chapter 7

Findlay, G. 1972. *Dr. Robert Broom, F.R.S.; palaeontologist and physician, 1866–1951: a biography, appreciation and bibliography*. A. A. Balkema, Amsterdam.

Kuljian, C. 2016. *Darwin's Hunch: Science Race and the Search for Human Origins*. Jacana Media, Johannesburg.

Merritt, J. F. 2010. *The Biology of Small Mammals*. Johns Hopkins University Press, Baltimore.

Wallace, D. R. 2004. *Beasts of Eden: Walking Whales, Dawn Horses, and Other Enigmas of Mammal Evolution*. University of California Press, Berkeley.

Chapter 8

Benoit, J., Manger, P. R. & Rubidge, B. S. 2016. Palaeoneurological clues to the evolution of defining mammalian soft tissue traits. *Scientific Reports* 6: 25604.

Cunningham, J. A., Rahman, I. A., Lautenschlager, S., Rayfield, E. J. & Donoghue, P. C. 2014. A virtual world of paleontology. *Trends in Ecology & Evolution* 29: 347–357.

Miller, H. 1858. *The Cruise of the Betsey, Or, A Summer Ramble Among the Fossiliferous Deposits of the Hebrides: With Rambles of a Geologist, Or, Ten Thousand Miles Over the Fossiliferous Deposits of Scotland.* Constable & Co, Edinburgh.

Oftedal, O. T. 2012. The evolution of milk secretion and its ancient origins. *Animal: an international journal of animal bioscience* 6: 355–368.

Chapter 9

Drew, L. 2017. *I, Mammal: The Story of What Makes Us Mammals.* Bloomsbury Sigma, London.

Panciroli, E. 4 July 2018. Beijing fossil exhibition prompts rethink of mammal evolution. *Guardian.*

Chapter 10

Cifelli, R. L. & Fostowicz-Frelik, Ł. 2016. Legacy of the Gobi Desert: Papers in Memory of Zofia Kielan-Jaworowska. *Acta Palaeontologica Polonica* 67.

Coiro, M., Doyle, J. A. & Hilton, J. 2019. How deep is the conflict between molecular and fossil evidence on the age of angiosperms? *New Phytologist* 223: 83–99.

Kielan-Jaworowska, Z. 1975. Late Cretaceous Mammals and Dinosaurs from the Gobi Desert: Fossils excavated by the Polish-Mongolian Paleontological Expeditions of 1963–71 cast new light on primitive mammals and dinosaurs and on faunal interchange between Asia and North America. *American Scientist* 63: 150–159.

Kielan-Jaworowska, Z. 2013. *In Pursuit of Early Mammals.* Indiana University Press, Bloomington.

Kühne, W. G. 1956. The Liassic Therapsid *Oligokyphus.* British Museum, London.

Mayor, A. 2000. *The First Fossil Hunters: Palaeontology in Greek and Roman Times.* Princeton University Press, Princeton New Jersey.

Chapter 11

Agustí, J. 2005. *Mammoths, Sabertooths, and Hominids: 65 Million Years of Mammalian Evolution in Europe.* Columbia University Press, New York.

Barnett, R. 2019. *The Missing Lynx: The Past and Future of Britain's Lost Mammals.* Bloomsbury Sigma, London.

Grossnickle, D. M., Smith, S. M. & Wilson, G. P. 2019. Untangling the multiple ecological radiations of early mammals. *Trends in Ecology and Evolution* 34: 936–949.

Prothero, D. 2019. *The Princeton Field Guide to Prehistoric Mammals* (Princeton Field Guides). Princeton University Press, Princeton.

Slater, G. J. 2013. Phylogenetic evidence for a shift in the mode of mammalian body size evolution at the Cretaceous–Palaeogene boundary. *Methods in Ecology and Evolution* 4: 734–744.

Zachos, F. & Asher, R. 2018. *Mammalian Evolution, Diversity and Systematics*. Walter de Gruyter GmbH & Co KG, Berlin.

Epilogue

Dixon, D. 2018. *After Man, A Zoology of the Future* (updated edition). Breakdown Press, London.

Francis, R. C. 2015. *Domesticated: Evolution in a Man-Made World*. W. W. Norton & Company, New York.

Pecl, G. T., Araújo, M. B., Bell, J. D., Blanchard, J., Bonebrake, T. C., Chen, I. C., Clark, T. D., Colwell, R. K., Danielsen, F., Evengård, B. & Falconi, L. 2017. Biodiversity redistribution under climate change: Impacts on ecosystems and human well-being. *Science* 355: eaai9214.

Weissman, A. 2008. *The World Without Us*. Virgin Books, London.

Index

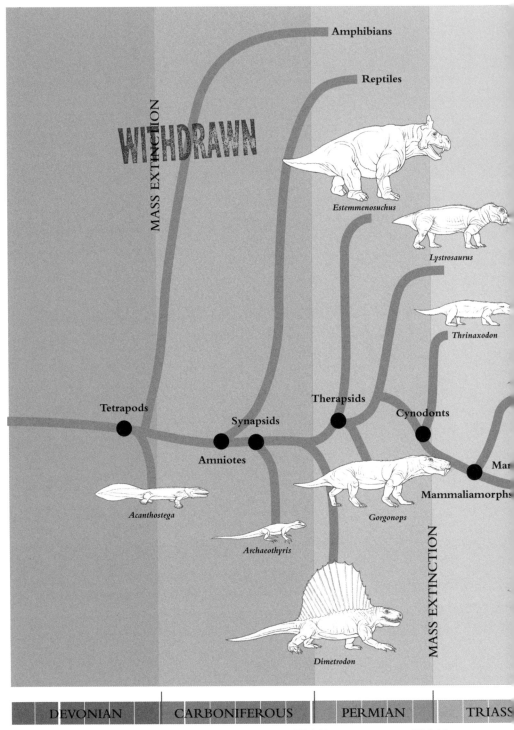

WITHDRAWN

MASS EXTINCTION

Amphibians

Reptiles

Estemmenosuchus

Lystrosaurus

Thrinaxodon

Tetrapods

Synapsids

Therapsids

Cynodonts

Amniotes

Mar

Mammaliamorphs

MASS EXTINCTION

Acanthostega

Archaeothyris

Gorgonops

Dimetrodon

| DEVONIAN | CARBONIFEROUS | PERMIAN | TRIASS |

419.2 Mya   358.9 Mya   298.9 Mya   251.9 Mya